LITTLE GREEN LIES

Little Green Lies

An Exposé of Twelve Environmental Myths

Jeff Bennett

Published in 2012 by Connor Court Publishing Pty Ltd.

Copyright © Jeff Bennett 2012

ALL RIGHTS RESERVED. This book contains material protected under International and Federal Copyright Laws and Treaties. Any unauthorized reprint or use of this material is prohibited. No part of this book may be reproduced or transmitted in any form or by any means, electronic or mechanical, including photocopying, recording, or by any information storage and retrieval system without express written permission from the publisher.

Connor Court Publishing Pty Ltd.
PO Box 1
Ballan VIC 3342
sales@connorcourt.com
www.connorcourt.com

ISBN: 9781921421648 (pbk.)

Cover design by Ian James

Typesetting by Chris Ulyatt

Printed in Australia

Contents

Acknowledgements   vii

Prologue   ix

1. Peak Oil   1
2. Renewable Energy   21
3. Index-based Consumption Decisions   43
4. Population   67
5. Trade and the Environment   89
6. Waste   109
7. Using Resources 'Efficiently'   129
8. The Infinitely Valuable Environment   145
9. Climate Change   163
10. Protecting the Environment, Privately   185
11. Agriculture and Mining   209
12. The Precautionary Principle   233

Index   249

Acknowledgements

This book was written while I was on sabbatical leave from my position at the Australian National University throughout 2011. During that year, I took the opportunity to visit numerous universities, research organisations and colleagues in New Zealand, Europe and North America. The experience exposed me to a rich array of issues and ideas that has been formative in the preparation of this book. I am very grateful to my hosts at these various venues for their support for my endeavour and to the ANU for granting me leave.

Each chapter of the book has been reviewed by two respected colleagues. Some agreed to look at two chapters. I am most appreciative of the effort these people made to improving both the content and the style of the book.

Final editing of the book was provided by Chris Ulyatt, along with his customary good cheer.

Jeff Bennett
March 2012

Prologue
Environmental fact or fiction

We live in the information age. We are constantly exposed to news and ideas through conventional and evolving media sources. Social networking sites and the blogosphere have ensured that information transmission is rapid and the volume overwhelming.

Different sources of information compete for our attention. The costs of dissemination have fallen to such unprecedented pittances that there are few remaining barriers to supply. The resultant competition between sources sounds worthwhile and indeed exciting. However, the task of processing it all and determining just what information is useful and what should be trashed is costly. It takes time. The barrage we experience means that there is little opportunity for quiet contemplation of the various pieces of information presented to us.

Such quiet contemplation is important because 'all that glitters is not gold'. We need to be able to identify any fiction that masquerades as fact. This appears to be the case with issues that involve the environment in particular. The information barrage confronting us that deals with the fate of the planet, and our role in determining it, is particularly overwhelming. Views are held vehemently and emotions run high when the stakes are purported to be as significant as the future survival of the species with which we co-habit this earth and the well-being of our children and their descendants.

The focus of this book is on a set of views, presented in the form of propositions about the way we view and manage the environment. These propositions often appear in the media. They are worrying because the implications that flow from them challenge the comfort of the *status quo*. Yet they are not universally accepted and debate surrounding them is often heated.

The goal here is to scrutinise the logic of these propositions so that their significance in directing environmental public policy is better understood and can be re-assessed.

The contention is that these propositions are indeed 'little green lies': That is, distortions of the truth delivered into the public debate on environmental issues in order to 'protect' people from the truth.

The analogy is to 'little white lies'. These are untruths for which we may be excused because telling the truth may hurt ourselves or those around us to the extent that a lie is warranted. For instance, I might be excused for telling my sister that I like the purple shirt she has kindly given to me for my birthday, even though I don't. Without the 'little white lie', my judgement is that my sister would be worse off. Furthermore, without the 'little white lie' I'm thinking that I would be worse off.

In the same way, people might promulgate 'little green lies' with a view to preventing what they see to be harm to other people. For instance, environmental advocates who advance the 'food miles' concept – a 'little green lie' considered in more detail in Chapter 3 – try to convince consumers that they should buy local produce rather than food grown in some far distant field. Their argument is that the local food is less environmentally harmful because it doesn't have to be transported as far and so involves less use of energy and hence less pollution. Even though this is not necessarily the case, 'food miles' devotees believe that they are preventing people from harming their environment and themselves if they can make sure that their 'little green lie' is believed.

'Little green lies' may also be good for their tellers.

If more people believe the food miles story and buy 'local', environmental advocates are better off because they see more people contributing to the achievement of the environmental goals that they hold dear.

However, a 'food miles' proponent may also be a local farmer whose produce will be in higher demand if the concept is accepted by consumers. The belief on the part of the consumer that local foods trump competing imported foods, despite a higher price, generates an improvement in local farm income and wealth.

If 'little green lies' are believed by the broader public, then their promulgators may also be able to secure public policy goals that further the satisfaction of their own preferences at the expense of others. For instance, environmental lobbyists may try to convince more and more members of the public of the 'little green lie' reviewed in Chapter 8 that the environment must not be harmed (as a moral imperative) because it is of infinite value. If they were successful, then the political process would start to generate more and more policy outcomes that protect the environment.

This makes people who have strong preferences for the environment happier and hence gives them a reason to tell the 'little green lie'.

But there are costs to such policies and therefore there are costs to the telling of 'little green lies'. For instance, if the political force generated by the 'little green lie' that the natural environment has infinite value is sufficiently strong, an additional area of forest environment may be protected by a ban on logging. This would mean that people would have to pay more for the timber products that would have otherwise been harvested from that forest. The higher price of timber is a cost – but one that is spread thinly across the whole of society. If spread thinly enough, then individually, people are more likely to be willing to bear their share of the costs – given their newly formed perceptions of just how important the environment is. A logging ban would also mean that some forest livelihoods would be lost. Some forest businesses may lose money. Some timber workers may lose their jobs. Those would be costs too. The higher that cost burden and the more concentrated it is on specific, politically influential groups in society, the less likely it is that the environmental protection policy options supported by the 'little green lie' will be accepted. If the affected group is politically weak or irrelevant, perhaps because their votes are not going to be affected by the change in policy or because they constitute a group within society that is largely unnoticed by other voters, the policy based on the 'little-green-lie' is more likely to be introduced.

The 'little white lies' analogy would suggest that it makes everyone better off if we just let 'little green lies' go past without correction. For instance, if my wife told my sister that I don't like the colour purple, and so reveal the truth about my dislike for the gifted shirt, then all parties are likely to be worse off.

This is where the analogy with 'little white lies' ends. 'Little green lies' are not harmless. The argument advanced in this book is that revealing the truth about 'little green lies' will make society as a whole better off.

Certainly, exposing the 'little green lies' and by doing so, preventing or reversing public and private decisions that flow from them would make their champions worse off. The special interests that are advanced by the acceptance of the 'little green lies' would be set back. However, the well-being of the members of the public who would otherwise bear the costs arising from the acceptance of the 'little green lies' would be improved.

The analogy with 'little white lies' is also on shaky ground in terms of whether or not the teller knows that it is a lie. When I thank my sister for her gift, I know that I don't like purple. In contrast, those who advance 'little green lies' may not be aware of their position's lack of veracity. Indeed they may be convinced that their advancing the 'cause' is for the greater good. For instance, consider an advocate of the renewable energy 'little green lie' – reviewed in Chapter 2. They may see the pollution associated with operating fossil fuel-sourced power stations and make the comparison between renewable and non-renewables. On that basis they may decide that, for themselves, avoiding the pollution cost of the non-renewable energy is worth paying the higher price of using renewables.

There are two specific problems associated with the process used to come to that conclusion. First, the individual's concerns about the damage caused by the non-renewable's pollution may not be shared by the majority of the population. Second the understanding of the impacts of choosing renewable energy sources over non-renewables may be limited to a single dimension – namely, the pollution associated with operating the power plants.

The goal in this book is to move away from this type of approach. The two specific problems associated with the renewable energy 'little green lie' can be generalised as two general flaws. First, the 'little green lie' approach is based on individual interests and so ignores the well-being of society as a whole. Second, it is focused on single issues and so misses the 'big-picture' consequences of the actions taken. These two flaws are now elaborated.

The first characteristic is problematic because a conclusion, and its policy outcome, drawn on the basis of the preferences of an individual or a specific interest group within society, may well be detrimental to society as a whole.

The little white lie analogy provides a useful comparative insight here. I know my own preference for colours. It is straightforward for me to decide that I don't like the purple shirt my sister has given me. Likewise, the 'little green lie' teller knows their own preferences for the environment. However, I can't tell what preferences other people may have for purple shirts. Nor can the 'little green lie' teller say with certainty how the rest of society thinks about the environment. Hence, just as my dislike for purple cannot be inferred to be a general community dislike for purple, nor can the 'little green lie' teller's love for the environment be held true for a majority of people in society.

I could try to convince the broader community that purple shirts are in poor taste, so that I am less affronted when I see one being worn.

This is what those promulgating 'little green lies' do when setting out to publicize their environmental viewpoint.

I could also try to convince enough people to support my political campaign to have purple shirts banned in the same way that politicians are lobbied by 'little green lie' tellers.

Neither of these strategies would be deemed to be acceptable behaviour for me by the rest of society. At best, I would be told to mind my own business. At worst, I'd be accused of trying to brainwash my fellow citizens. By imposing my preferences on others, my gain is overwhelmed by the aggregation of the losses endured by those who do not share my preferences.

These retorts are infrequently levelled at 'little green lie' tellers.

The second characteristic is problematic because the concentration on a single dimension of an issue can deliver perverse outcomes. Society and the environment are examples of complex interdependent systems. Key features of such systems are the feedback loops that can accentuate change or counteract it. The implications of change are rarely straightforward and to concentrate on a single dimension of change is likely to be inadequate.

For instance, the population cap 'little green lie' – the subject of Chapter 4 – is focused on the pollution impacts of more people and the demands they place on scarce resources. It ignores the capacity of people to solve problems they may have created and to avoid foreseen problems. Economic growth and trade – examined in Chapter 5 – likewise involve multiple dimensions. Some of these may be detrimental to the environment but others may provide incentives and the capacity to improve environmental conditions. Drawing conclusions on the basis of just one side of the argument is short-sighted and potentially harmful to society.

The logical strategy to avoid these problematic characteristics of 'little green lies' is to adopt an analytical stance that is based on a societal perspective. That involves looking beyond the immediacy of an environmental issue. It means taking into account the many dimensions across time and space that characterise the interaction between society and the environment.

There is little doubt that this approach will be contentious. Taking a 'societal' perspective necessitates an understanding of society's preferences. They are innately difficult to observe. The difficulties are especially vexatious where the environment is concerned and there are only limited windows through which these preferences can be observed. For society, we can generally look at people's behaviour in what they buy and sell to infer the strength of their preferences. But for environmental goods and services, there are few markets from which we can infer community-wide preferences. Differentiating between truth and lies becomes difficult.

Furthermore, many 'little green lies' revolve around what may or may not happen in the future. For instance, the global warming (or more recently, global climate change) 'little green lie' – considered in Chapter 9 – is characterised by opposing groups of climate scientists (some claiming to represent a consensus view) forecasting the fate of the planet over coming centuries. There is much uncertainty surrounding this and other issues that are the focus of this book.

Put simply, the truth may be out there – but it's very hard to pin down.

Some purveyors of 'little green lies' will reject strenuously the suggestion that the truth of these propositions is in question. Indeed 'truth' is not necessarily an absolute. Its definition is made from the position of those making the determination. And in many of the cases considered in this book, the people involved in the debates come from remarkably different standpoints.

Furthermore those who tell the 'little green lies' may reject the claim that they have vested interests and suggest, alternatively, that the motivation for writing this book is a vested interest in seeing the environment damaged in order to secure financial advantage.

A further goal of this book is, therefore, to demonstrate that many of the 'little green lies', if taken to their logical policy conclusions, could be counterproductive even for the environment. They are argued to be potentially 'counter-intuitive' in terms of their environmental consequences because of their inappropriate focus on just one dimension of the issue at hand.

In a nutshell

There are twelve propositions addressed in the twelve chapters of this book. Although each proposition is considered in a separate chapter, many of them are interrelated. In the list of the propositions that follows, a short outline of each 'little green lie' is set out along with a brief exposition of the counter-proposition that will be advanced in this volume.

Proposition 1: 'Peak Oil' has been reached.

The annual production of oil, while rising over the last century, has levelled off and is about to fall because of growing scarcity. Such is our dependence on oil and the fast rate at which we are using it that we now need to take active policy measures to save what we have left.

BUT

No-one knows for sure what petroleum reserves are available. As known reserves are depleted, price rises stimulate more exploration and technological advances that will expand the available supply of petroleum as well as substitute energy sources. Price rises will also help to restrain oil consumption and re-channel energy demand to substitutes.

Proposition 2: Renewable energy production should be stimulated.

Non-renewable energy supplies are being depleted so quickly that we will soon experience power shortages. Non-renewables are also 'dirty' sources of energy. Renewable energy production must be stimulated to ensure the on-going supply of clean energy.

BUT

Renewable energy sources are limited in their short to medium term potential to meet demand. Picking the 'winners' to be stimulated is likely to be mistaken with rapidly evolving technological change. Renewables have their own environmental downsides.

Proposition 3: Consumption choices need to be informed by products' environmental characteristics such as their 'food miles', 'ecological footprint', 'embodied energy', 'virtual water' and 'carbon footprint'.

People need to be aware of the impacts they have on energy, the ecology, water, climate, etc. when they buy goods and services so that they can reduce their impact on those resource. Each resource is scarce and valuable. We need to conserve them, especially for future generations.

BUT

By focusing on just one scarce resource (water, energy, etc.) in their consumption decisions, people can ignore their impacts on other scarce resources. A 'false economy' results. When impacts on multiple resources are combined in an index, more distortions arise. Prices do a better job of signalling resource scarcity.

Proposition 4: World population should be capped.

More people mean more pressure on the world's scarce resources, including the environment. The only way to protect the environment, stop starvation and to ensure that there are enough resources for future generations is to stop population growth.

BUT

People are a resource. They have the capability to develop innovative technologies and institutions to deal with growing scarcity in specific resources. New ways to satisfy people's wants and new sources of scarce resources can be discovered. If limits are installed, who will they affect?

Proposition 5: Economic growth and trade are bad for the environment.

Economic growth, fuelled by international trade, means more pressure on scarce resources, including the environment. To protect the environment and to save resources for future generations, trade should be restricted to cut growth.

BUT

Trade and growth bring wealth to people. Wealth increases peoples' demands for environmental protection and the ability of society to provide environmental protection, especially through technological development.

Proposition 6: No waste should go to landfill.

Waste should not be 'wasted'. It is a resource that can be re-used and re-cycled. Sending waste to landfill means that more 'virgin' resources must be harvested/mined. Waste in landfill can also be a source of air and water pollution.

BUT

Recycling and re-using 'waste' is a process that uses scarce resources. Policies that prevent landfill disposal can cause more resources to be used than they save and do not necessarily reduce virgin resource use. Landfills need not be pollution sources.

Proposition 7: Resources such as water and energy should be used 'efficiently', whatever it costs.

Resources such as water and energy are scarce. Use of these resources needs to be minimised so that future generations will have enough. Governments should invest in technologies that will ensure the least amount of energy and water is used in producing goods and services.

BUT

Investing in 'efficiency' measures often means using other scarce resources as substitutes for energy and water. A 'false economy' results because the other resources, including labour and capital, may well be scarcer than energy and water.

Proposition 8: The environment is of infinite value and must not be harmed.

The environment provides us with our 'life-support-system'. Without it we cannot survive and so we should protect it at all costs.

BUT

Without the environment we could not exist and so its absolute value is infinite. However, that is not the relevant question for policy. Changes to the state of the environment yield finite benefits and costs that need to be traded off when making policy decisions.

Proposition 9: We must reduce greenhouse gas (GHG) emissions to avoid global climate change.

Human-induced global climate change is a serious threat to the continued ability of the planet to support humanity and current ecosystems. The damage caused by climate change will be so large that GHG emissions must be reduced now.

BUT

Reducing GHG emissions would be costly. The decision to bear those costs should be made with reference to the expected benefits that reduced GHG emissions would provide. Reducing GHG emissions will not eliminate the risk of climate change.

Proposition 10: The care of the environment cannot be entrusted to the private sector.

The environment provides 'public goods' that should be available to all free of charge. That means the government has to be responsible for caring for the environment. The private sector will either destroy it or try to profit from it.

BUT

The public sector will face problems managing the environment. Gathering information for effective decision-making is costly. Politicians' and bureaucrats' incentives can conflict with the public's best interest. Private solutions can be lower cost and better aligned with what the community wants.

Proposition 11: Agriculture and mining are always in conflict with the environment.

Agriculture and mining are extractive industries which deplete our stock of natural resources, often irreversibly. They also cause environmental degradation including soil erosion, biodiversity loss and contamination of the gene pool, water and air.

BUT

While there are some trade-offs between agriculture, mining and the environment, these can be reduced through the use of management techniques and technologies. Offsets and remediation work on farms and mines can maintain and even improve the condition of the environment.

Proposition 12: Decisions regarding the future of the environment should be made using the 'Precautionary Principle'.

If there is a risk that a proposed action will harm the environment, the Precautionary Principle requires policy makers to regulate against that harm and to place the burden of proof on those proposing an action that it will not cause environmental damage.

BUT

There is always some risk of environmental harm resulting from human action. Demonstrating that there is no risk of harm is impossible. There are also uncertainties associated with not taking action which the Precautionary Principle ignores.

These propositions have been selected in a process that reflects a combination of the author's interests and the frequency with which the 'little green lie' crops up in popular and even scholarly debate. The list of 12 propositions is not exclusive. However, it is of interest that the list is not completely divergent from the list of chapters that featured in an edited volume addressing the same overall concerns that was published 17 years ago.[1] Since the publication of *Tall Green Tales* it would appear that the preponderance of 'little green lies' has not diminished and that the predominant 'little green lies' have also remained fairly static. Nor has the ferocity of the debate that often accompanies them diminished. The 'tunnel vision' that is embodied in many of them – that focus on one particular (environmental) aspect of an issue to the exclusion of all other aspects that are often countervailing – seems to be an on-going feature of the environmental debate. It is certainly time that the 'blinkers' were removed once more and the 'little green lies' were re-addressed.

---

[1] Bennett, J. (ed.) (1995), *Tall Green Tales*, Institute of Public Affairs: Perth.

1. Peak Oil

Proposition: 'Peak Oil' has been reached.

The annual production of oil, while rising over the last century, has levelled off and is about to fall because of growing scarcity. Such is our dependence on oil and the fast rate at which we are using it that we now need to take active policy measures to save what we have left.

BUT

No-one knows for sure what petroleum reserves are available. As known reserves are depleted, price rises stimulate more exploration and technological advances that will expand the available supply of petroleum as well as substitute energy sources. Price rises will also help to restrain oil consumption and re-channel energy demand to substitutes.

The scarcity of oil

We all know how important oil is to us – individually, nationally and globally. We use it every day in a host of ways, either directly or indirectly.

As a source of fuel it powers the cars, buses and planes that we rely on to take us to work, to shop, to visit friends and relatives and on holidays. It fuels the ships and trucks that transport our food, our clothes, and our shelter from sources of supply both far and near. We burn it and its derivatives to stay warm through cold winters.

As a raw material, oil plays a critical role in the production of the goods and services we rely on to sustain our standard of living. It is used for the production of fertilisers, pesticides and herbicides on which the modern farming systems that produce our food supplies and natural fibres are dependent. Oil is also used in the production of most plastics products and the synthetic fabrics that we find so useful.

It's hard to imagine a world without oil.

So it is no wonder that when we hear news that the world's oil 'fuel gauge' is about to enter the red zone, we begin to think twice about our futures. We become especially anxious when we realise that the rate of oil usage has been increasingly rapidly with the emergence of countries such as India and China as nations of car owners.

It's a situation akin to the motorist taking a look at a falling petrol gauge, knowing that the road ahead is a steep climb that will use a lot of fuel: the search for a place to re-fuel starts to become urgent. The thought of being stuck out in the middle of nowhere with an empty tank is a powerful incentive to make sure of a good supply.

A striking example of this sort of thinking is to be found on the website [www.worldometers.info]. By viewing this site you can see at any time the (estimated) number of barrels of oil remaining in the world and the (estimated) number of days we have left where we can drive up to the petrol pump. It was 15,577 days today. That's around 42 years.

This is the essence of the peak oil hypothesis. It involves a consideration of the relationship between estimates, over time, of the stock of oil reserves still to be extracted and the rate of change in the demand for oil. Once increases in demand can't be matched by supply expansions, and annual output starts to fall, peak oil is said to have been reached. The implication, too, is that the remaining supply of oil is going to be used up much faster than before 'peak oil' – simply because our rate of use is now so much higher.

Peak oil is, however, more than a quantitative concept. It forms the basis for a number of policy positions that revolve around two goals.

First, policies seek out ways to slow down the consumption of oil so as to postpone the eventuality of the 'dry tank'. Second, they aim to develop alternative ways of providing all the goods and services for which we are currently oil-dependent: This is the search for a new 'fuel stop'.

This chapter is therefore focused around these two peak oil questions:[1]

- Are we close to reaching peak oil?; and,
- If so, what should we do about it?

As such, the goal of the chapter is to assess if peak oil is a 'little green lie' and whether or not the policies that are purported to be consequences of peak oil are well justified. Are we at the point where annual output is about to start falling? If so, does it matter? If it does, what should we (in a public policy sense) do?

Are we there yet?

By looking over the statistics about world oil consumption, it is comparatively straightforward to determine if we have reached peak oil. The evidence is that the trend over recent years is for oil consumption – and hence the production of oil – to be rising.[2] In 1965 the world used around 31,000 barrels of oil per day. By 1980 that had risen to over 63,000 barrels and, by 2009, 84,077 barrels of crude were being used each day. The trend in use is mixed across different countries. In the USA, for instance, the rate of use rose from 11,522 barrels a day in 1965 to 18,686 in 2009; whereas in China, over the same period, the increase has been from 217 to 8,625 barrels per day.

---

1   Peak oil can be considered in absolute terms (the amount of oil being consumed per day) or as a relative concept (the days of oil consumption that remain). The two are clearly related: The more that is consumed per day the implication is that fewer days of oil consumption remain. While both aspects are considered in this chapter, more emphasis is given to the relative concept because of its implications for the scarcity of the resource and the policies that arise from those implications.
2   www.eia.gov/emeu/international/RecentPetroleumConsumptionBarrelsperDay.xls and http://www.bp.com/subsection.do?categoryId=9023761&contentId=7044545

However, this evidence to dismiss 'peak oil' is not 100 per cent convincing. For instance, in 2009, the daily rate of oil consumption fell relative to 2008. This amounted to a 1.7 per cent annual reduction. And it came on top of a fall from 2007 to 2008 of around 1 per cent. Over this period, significant declines in oil consumption were experienced in the OECD nations while emerging economies' usages continued to increase. Were these two years' statistics enough to conclude that peak oil has arrived?

Understanding the answer to that question and, more generally, understanding the world's proximity to peak oil production involves understanding the forces that drive oil supply and demand trends now and into the future. This is a vexed exercise that many have attempted and many have – with the wisdom afforded by hindsight – failed. In a sense, the failures are not surprising. So many factors influence oil production that to predict future trends with complete accuracy is impossible. Settling the question of whether or not the world is about to reach 'peak oil' is therefore also an impossible task.

For instance, consider the drop in oil consumption from 2007 through to 2009. Was that due to a decrease in overall supply availability (and thus indicative of 'peak oil')? Or was it driven by a drop in demand from the OECD countries due to the global financial crisis that hit economic growth in most of the developed world?

Notwithstanding these complexities, what can be done to assess the 'peak oil' proposition is an analysis of the broad trends in some of the factors that influence the rate of oil use.

The driving forces

The first point of interest in such an analysis is the state of known oil reserves. If 'peak oil' were imminent, then we would logically expect to see a run-down of known reserves. Even if current consumption was not declining to indicate 'peak oil', the available stock would be falling.

The figures about reserves that are provided by BP[3] are in the form of a 'reserves-to-production' (R/P) ratio. This ratio can be interpreted as the number of years that the currently known reserves will be able to support present levels of production (and hence consumption). It is the same ratio as that reported so graphically by 'worldometers' as the years until the end of oil. Over the period 1985 to 2009, the R/P ratio has increased from around 38 years to 45.7 years and the trend over those years has been of relative stability around the low-40 year mark. With oil consumption constantly drawing down reserves and with pressures from emerging economies to extract at an increasing rate, how could the time until we run out of oil be staying still? Oil's supposed status as a non-renewable resource looks from these data to be under challenge!

The reason for the stability in R/P ratio is 'given away' by BP[4] in a footnote to these data:

> Global proved oil reserves rose by 0.7 billion barrels to 1,333.1 billion barrels, with an R/P ratio of 45.7 years. Increases in Indonesia and Saudi Arabia have more than offset declines in Norway, Mexico and Vietnam. The 2008 figure has been revised higher by 74.4 billion barrels, largely due to an increase in Venezuelan official reserves. (p. 10)

Crude oil is a finite, non-renewable resource. The rate at which the resource is created by nature is small compared to the rate at which the resource is being used. However, the amount of oil that is available for future use is essentially unknown. It is not a geologically given amount. It changes over time. For instance, the stock of oil can increase as more information on reserves is gathered through exploration. With more

---

3   http://www.bp.com/subsection.do?categoryId=9023761&contentId=7044545
4   http://www.bp.com/liveassets/bp_internet/globalbp/globalbp_uk_english/reports_and_publications/statistical_energy_review_2008/STAGING/local_assets/2010_downloads/oil_section_2010.pdf

effort devoted to exploration, it is likely that more reserves will be found. That is, the top line in the R/P ratio (R) is not a constant.

It is not only through exploration that reserves can grow. Technological change can create expansions in reserves even from geological features that are comparatively well-explored. Oil reserves are not like tanks from which 100 per cent of the crude can be pumped. Although the oil from some wells is pressurised, at least in the initial phases of extraction operations,[5] no well is able to extract all the oil from the geological features in which the oil is deposited. The percentage of oil extracted – the industry standard is currently around 40 per cent[6] – is a matter of intense technical research and development, the results of which can have significant impacts on the overall extent of the known stock of oil.

But reserves are not only questions of geology and technology. They are also dependent on economic forces. An oil deposit is only a reserve if it is financially feasible for the resource to be extracted. If the price of oil goes up or the costs of finding and drilling for oil go down, then the profitability of oil exploration and recovery increases. With increasing profitability, more oil reserves will come into being. This is not because the resource miraculously becomes 'renewable'. Rather, it is because previously unprofitable wells in known geological features will then be counted as part of the stock. Furthermore, more exploration in areas previously thought to be too costly to develop (deeper water, more extreme climates, more environmentally sensitive) will be initiated and eventually add to total reserves.

The R/P ratio is, of course, also affected by levels of production. This in turn is dependent on the extent of demand for oil. As already

---

5   Extraction from these wells provide the engineering challenges associated with controlling the extent of flows. The Deepwater Horizon well in the Gulf of Mexico illustrated these challenges in April 2010 when the well's blow-out protector valve failed.
6   Green D. and G. Willhite (2003), *Enhanced Oil Recovery*, Society of Petroleum Engineers Textbook Series, Vol. 6, Richardson, Texas.

indicated, the strength of the world economy has a role to play in determining production: If growth in the world's key economies is weak (or even negative as in the years of the global financial crisis), then people use less oil as the production and consumption of the full spectrum of goods and services declines. But many other factors also impact on demand.

Just as the price of oil is a key driver of supply, so too does it have a key role in determining the amount of oil that households, businesses and governments consume. In the fundamentally market-driven world economy, price plays a key role as the regulator of the demand for oil. As the price of crude oil rises, the amount people want to buy falls – albeit with a lag, as adjustments take place. Short-run price jumps may not have strong immediate impacts, but as a price-hike stays put for longer, more and more people make adjustments to cut their consumption. This can happen in all sorts of different ways.

First, people – households, businesses and governments – may start by simply cutting back their consumption by making do with less. Then they may explore ways in which they can still keep using oil but achieving their objectives with less of it. For instance, an investment in modernising the technology that uses oil in a production process (purchasing more advanced insulation for an oil fired furnace) or a consumption activity (travel using a new, more efficient, but still petrol-powered car) may become worthwhile under a consistently higher price. Perhaps most significantly, oil users will start to investigate alternative means to supply their energy requirements. This could mean power generators switching from oil-fired thermal plants to coal or gas systems. Commuters may decide to take a bus or train to work instead of their cars, or they may convert their cars to run on compressed natural gas (CNG) or liquefied petroleum gas (LPG).

These changes in turn trigger responses. With more people looking for alternatives, suppliers of potential oil substitutes have a competitive incentive to provide options at the lowest financially viable cost.

Innovations in alternatives are stimulated. Oil use efficiency, too, becomes a focus of research and development.

In all these ways, the level of P in the R/P ratio is moderated as prices rise and hence the ratio is increased, spinning out the years of oil still remaining.

Oils that ain't oils?

Perhaps most significant on the list of factors that are influenced by the price of oil is the stimulation of the development and the production of energy alternatives. A close look at some of the alternatives that are being developed and produced will lead us to further question the notion of 'peak oil'. Does the 'peak oil' concept have any policy relevance?

If the alternatives to oil being brought forward are providing exactly the same function that oil has provided, then should they be included in the R/P ratio as a part of 'R'? For instance, with the price of oil topping the USD100 mark, the production of oil from the tar sands deposits in Canada delivers a sound financial reward to their owners. Should those reserves be added to 'R'? Essentially, oil derived from tar sands can be refined to produce the same compounds as crude oil, albeit in different proportions. Hence, diesel fuel distilled from tar sands is no different from diesel fuel distilled from crude oil. Its inclusion within 'R' seems logically sound.

So, too, the shale and coal seam gas deposits in the United States and Australia are financially worthwhile to extract as energy demands shift away from the relatively more highly priced crude oil. Yes, gas is not oil, but it is still a fossil fuel and is frequently found in association with oil. Furthermore, it is so closely able to perform many of the same functions as oil that it would seem strange to differentiate the two. According to 'worldometers', there are 61,002 days left of gas production (given current consumption rates). And more reserves are being proven every day. That would be a significant addition (around

167 years) to the 42 years of oil.

Then there is coal. Coal is a source of energy just as is oil. It has taken the place of oil in many applications, especially electricity generation, over the last 50 years as their relative prices have shifted in favour of coal. It is also possible, technically, to liquefy coal. In that form it is an even closer substitute for oil. Yes, coal is not oil but it can be just like it. With 'worldometers' tracking a total of 152,175 days' worth of coal production remaining from current reserves, adding that (around 500 years) to the oil R/P ratio again would make 'peak oil' look a long way off.

What these calculations make abundantly clear is that we've still got a long time yet before the fuel warning light comes on.

The price of oil

The discussion to this point has raised the key role that price plays in determining the fate of the 'peak oil' hypothesis. It is worth delving a little deeper into the story that the price of oil tells.

The first point to acknowledge is that the price of oil cannot be regarded as a paragon of market virtue. Manipulation of the price of oil and its derivatives by monopoly interests, cartels and monopsonists[7] (a single buyer), be they government or corporate, is rife. The Organisation of Petroleum Exporting Countries (OPEC) has historically played a leading role in manipulating world supply (at least in the short to medium term) attempting to advantage its member nations' interests.

In addition, market forces for oil are highly volatile and prices display strong fluctuations in response. Middle Eastern politics remain a particularly constant source of supply variability and hence price volatility: Even rumours of production disruption can cause spikes in

---

7   A monopoly exists when there is just one seller in a market. When there is just one buyer in a market a monopsony is said to occur. In both cases, market power is concentrated in the hands of one entity and the price is distorted. A cartel involves a group of suppliers banding together and acting as though they were a single supplier.

demand as buyers seek to add to their reserves of oil to buffer against uncertainty. That is not to downplay the importance of weather (such as cyclones in the Gulf of Mexico), accidents (including environmentally damaging shipping spills and off-shore well blow-outs), nationalisation strategies (for example, in Venezuela under President Chavez), and pipeline and exploration policies (in the USA), etc.

Nonetheless, price trends do tell a story that helps us to develop an understanding of the real issue that underpins concerns about 'peak oil': The overall relative scarcity of oil. In its simplest form, the argument would be that if oil is running out, as the 'peak oil' hypothesis maintains, then the price of oil should be rising in real terms[8] over time.

This is the fundamental conclusion drawn by the influential economist Harold Hotelling in his deliberations on the most efficient way that society should use its limited supply of a non-renewable (exhaustible) resource. In its simplest form, Hotelling showed that to ration efficiently the use of an ever-diminishing supply of a non-renewable resource amongst a constant set of demands, its price should rise by the amount of the interest rate each year.[9] If prices are not observed to behave in this manner, then we could expect that circumstances are different from those that Hotelling put together as his 'base case'. For instance, if the demand for oil was not constant, but rather was growing each year, then the price would be expected to rise faster through time. But if the stock of the resource was being augmented by further discoveries, the price rise over time would be lower or even negative.

What, then, do actual price trends tell us about the depletion of the oil resource? Over time, there has been a mixture of signals coming from observed price trends.[10] For the 100 years from 1870 through to 1970,

---

8 This is the price of oil corrected for any inflation in overall price levels.
9 Hotelling, H. (1931), 'The Economics of Exhaustible Resources', *Journal of Political Economy*, **39** (2): 137–175.
10 http://www.wtrg.com/prices.htm

the 'real' world price of oil – that is, the observed price corrected for inflation – fluctuated within a relatively narrow band between USD10 and USD20 per barrel (in 2008 prices). This pattern was broken by the first 'oil shock' of the late 1970s/early 1980s when a sequence of Middle Eastern crises (the Yon Kippur War and the Iran–Iraq war) pushed the world price up to around USD70 per barrel in 1980 before it fell back to fluctuate around USD30 per barrel between 1985 and 2003. A further price spike was then experienced as Chinese and Indian demand growth (and expectations of even more) coincided with more Middle Eastern unrest, this time in Kuwait and then Iraq. The real price rose to almost USD100 per barrel before falling back to the USD40 range when the global financial crisis-induced recession started to hit demand. The respite didn't last long, as uprisings in Egypt, Tunisia, Libya and Syria caused increased supply uncertainties as the world economy began to recover. Real prices again rose to around the USD100 per barrel mark.

The overwhelming message sent by price movements relates to the significance of Middle Eastern politics as a driver of price. This does demonstrate the importance of that region as a key supplier of crude oil, but the overall message of price is not one of impending resource exhaustion. Yes, real prices have, in general, risen: Median real oil prices have doubled from their pre-1980's levels to what they have been subsequently. However, this trend would be expected given the expansion of demand for the resource and the more costly processes (for instance, deeper off-shore wells) required to satisfy that growth.

The trend also shows the importance of 'choke prices' for the crude oil industry. The 'choke price' is the cost of supplying the best available substitute for crude oil, be it gas or tar sands-derived oil or a host of other possibilities. Once the price of crude hits the choke price, it sends a strong signal for buyers to switch – and for the sellers of the alternative sources of energy to ramp up production. Prices for crude around the USD100 per barrel mark are therefore unlikely to be sustained as supply-switching will become prominent.

Choke prices are not constant. Suppliers of alternatives have a strong incentive to lower their production costs so that energy users will switch to their product at even lower crude oil prices. This incentive applies not only to alternative fossil fuel energy suppliers but also to renewable power generators. Research and development of solar panels, wind generators, wave and tidal plants and geothermal stations aim to lower their per-unit costs of power generated. So too do nuclear power generators seek to lower their unit costs, including those associated with improving the safety performance of reactors. Manufacturers of products that use crude oil also try to improve their rates of efficiency. Car manufacturers, for example, strive to improve the fuel economy of their petrol- or diesel-driven engines. Hence the choke price is likely to come down as research and development efforts bear fruit.

Of course, owners of crude oil resources know all of this and take precautionary actions. OPEC, for instance, has made frequent attempts to manage the price of oil so that it neither dampens economic growth to the point where oil demand is severely constrained or where the producers of alternatives are given excessive encouragement. The ultimate danger of high prices for oil reserve holders is having stocks remaining that have been usurped by price-undercutting substitutes.

Policy responses
The prospect of a world without oil is daunting. We are heavily dependent on oil in almost all elements of our daily lives. Even the process of market exchange, especially international trade, that has been such a key part of the wealth generation witnessed over the past 200 years has been facilitated by the relatively cheap transportation afforded by oil. The notion of 'peak oil' is therefore one that touches a sensitive social nerve. This is particularly the case when the most readily perceived indicator of world crude oil price – the price of vehicle fuel at the bowser – keeps hitting levels never experienced before.

In response to this set of circumstances, it is not surprising that people will respond favourably to political activism calling for a governmental response. The threat of having to deal with oil shortages and its consequences in terms of losing all sorts of things we take for granted is sufficient to generate support.

Governments react in all sorts of different ways. One policy response is to impose restrictions on oil use. The rationale is to spread out available supplies over a longer time horizon. This reaction was common after the first oil price 'shock' of the late 1970s: Only cars with odd numbered registration plates were allowed to be on the road or to fill up with fuel on odd numbered calendar days. Fuel efficiency requirements in automobiles were imposed by legislation.

Of course the implication of such restrictions is that price is not doing a 'sufficient' job in rationing out available supplies. Why are prices not able to hold back demand so that these extra rationing schemes are needed? In many cases this was because the extra rationing schemes were introduced alongside measures to hold the price of oil-derived products down to levels that were politically acceptable. While we may all feel worried that oil is going to run out, we also worry when we have to pay a lot more for what is available!

The irony of price controls is that they require the use of these other means of restraining demand to the level of available supply. Restrictions on usage are only one form of such 'rationing' devices. Another common form is 'allocation by ordeal': Those of us who have queued for long periods in unpleasant circumstances to buy things such as tickets to high-profile sporting events or newly released electronic gadgets that are in short supply will know the meaning of this expression. The alternative forms of rationing don't mean that the goods are cheaper than they would be if price were allowed to do the rationing. It's just that their costs of purchase are not necessarily all monetary. The time spent in a queue is costly in terms of the other activities that could have been done during that time but were missed. The frustration of queuing may have a cost in terms of mental health! Those non-financial costs may be monetised by some people who have a high value on their time. They may find it cheaper to pay someone else to queue for them.

These non-price rationing systems are inevitably less efficient than allocation by price. If they are strictly enforced, they mean that buyers with relatively lower values for accessing the scarce resource can secure the resource in front of those with relatively higher values. People who have relatively low values for time may secure the resource in front of those who have high values for the resource itself. The outcome is an allocation that does not achieve the highest value for the available resource. That is the economist's definition of 'inefficient'. More value could be achieved from the resources that are available by using prices as the way to allocate resources instead of the alternatives.

On top of these allocative inefficiencies are the extra costs that have to be borne by the community to administer the scheme. A bureaucracy will need to be established to set up the rationing/price control scheme. More costs will be involved in monitoring and policing the scheme.

Perhaps most importantly, the use restrictions plus non-price allocation process don't really help to ensure a longer time-frame in which oil is available. All that is achieved is a rearrangement of who gains access to the available resource. The rate of use doesn't change, it's simply less efficient. In the process, all of the administrative costs involved means that other scarce resources that could have been used for achieving other goals are being wasted.

An even grimmer picture emerges if the threat of higher oil prices generates political support not only for controlled lower prices but also for subsidised production. This can occur when the artificially imposed lower prices cause the demand for oil to increase so that it out-strips available supplies. That means shortages. If the political reaction to these shortages of supply are very strong, governments may introduce production subsidies in an effort to increase the amount of oil available. Many, particularly oil-rich developing countries, have adopted this policy position. Indonesia has sought to increase the (subsidised) price of petrol, but after increases in 1998 were held at least partially responsible for the fall of the Suharto government, the strategy is no longer considered politically attractive. However, without reform, the fuel crisis is perpetuated and other policy mechanisms are required to deal with it indirectly. In December 2010, Iranian motorists were faced with petrol prices rising to 4,000 rials (40 US cents) from 1,000 rials (10 US cents) per litre. Such price levels were still well below prices paid by US motorists, and a 60 litre monthly quota was necessary to restrict demand. To fill a 60-litre fuel tank in Venezuela costs around USD1. Put simply, once price controls are in place, it's very hard to get rid of them. The analogy is with holding a tiger by its tail: There's no way of letting go without being hurt.

In such cases, price is not permitted to perform its rationing task. In addition, the production subsidies that are paid cause the rate of oil extraction to be accelerated. The outcome is resource exhaustion that is 'premature' unless stringent quota restrictions on use are enforced. And because the price has been kept low, there would be no process to give suppliers of alternative energy sources the incentive to innovate. Nor do consumers have the incentive to search for alternatives. The two actions of the price 'scissors' – the effect on supply and the effect on demand – are thus both curtailed. The 'end of oil' would come as a sudden shock, with no transition to a world of alternative sources. This truly would leave the world in a disastrous position.

So if policies designed to 'spread out' the available oil resource are so fraught with the danger of distorting both supply and demand conditions, what about trying to stimulate the development of alternative supplies of energy? This is the policy route being taken by many developed and even by some developing countries. The stimulus to develop alternative energy sources provided by a rising oil price is taken to be too weak and so government action is deemed (politically at least) to be necessary to speed things along. Subsidies, including tax concessions, for the production of alternative energy sources, chosen by government officials as being the 'best bets', are prominent in this category. Also included are research and development grants for alternatives.

The next chapter in this book is devoted to an analysis of these policy measures. Suffice it to say here that significant concerns arise from the implementation of such policies without them being able to claim much by way of an improvement in the capacity of the world to meet its energy demands.

A 'little green lie'?

There is no doubt that the world's supply of conventionally extracted crude oil is finite. Conceptually, there will come a time when production rates will start to decline and 'peak oil' will have been reached. However, there is little point to trying to determine that time. At best, the concept of 'peak oil' is illusory.

For a start, it is unclear whether or not we are close to reaching 'peak oil', even when the oil resource is defined strictly in terms of conventionally extracted crude oil. Adding in oil substitutes to the estimation of 'peak oil' adds further to that lack of clarity.

What is clear, however, is that the concept itself is of little practical use as an input into the making of oil policy. For a start, the notion of oil reserves is so hard to tie down. Reserves are a function not only of geology but also of economics. The latter can have significant and dynamic impacts on defining the size of reserves, and even the definition of what constitutes 'oil'. Similarly, the forces determining rates of production are variable as demand shifts through time. For example a technological breakthrough may be made in the development of a oil substitute that costs less than the current price of oil. That would mean any thought of 'peak oil' would vanish or become a mere curiosity – much like 'peak horses' became meaningless with the advent of the motor vehicle.

So 'peak oil' as a concept is of little practical use. Furthermore, it is certainly a dramatic oversimplification of the complexities facing the world in working through the coming decades of transition from oil to other sources of energy.

Is 'peak oil' then a case of a 'little green lie' built on a misunderstanding of these complexities or a more deliberate exercise in rent-seeking behaviour? The answer is probably a bit of both.

Evidence of a lack of understanding of the role that price plays in rationing scarce resource use can be found in numerous 'doomsayer' publications ranging back to Paul Ehrlich's treatise[11] and the pronouncements of the Club of Rome.[12] The process of taking the known reserves of a resource, dividing it by the current rate of use and predicting forthcoming catastrophic doom is well developed, even if naïve. Julian Simon's[13] efforts to dispel such myths have not been 100 per cent successful. The critical omission by Ehrlich and the Club of Rome of the role of resource substitution – and the role of price in stimulating substitutes – is an example of a 'little green lie' arising because of a blinkered view of the way in which the world works.

This is, then, the case of 'peak oil' proponents wanting to protect themselves and others from what they see, erroneously, as a forthcoming catastrophe.

It may not even be a case of these advocates of the 'little green lie' misunderstanding the principles behind the operation of price as a means of allocating scarce resources, but rather a demonstration of a lack of belief that those market principles will work in practice. Inherent to the arguments put in this chapter – that price signals will stimulate buyers to search for substitutes and for sellers to develop them – is the role of technological change: People will generate new ways to meet the needs that are currently being met by oil. Those who are pessimistic

---

11 Ehrlich, P. (1968), *The Population Bomb*, Ballantine Books, New York.
12 http://www.clubofrome.org/
13 http://www.juliansimon.com/

about the prospects of technological change to take on this challenge are thus most likely to be concerned to protect themselves and others from the threat of oil prices rising higher and higher without any alternatives coming into play. Policy measures that force others to cut back their consumption are seen by these technological pessimists as a way of dealing with the troubled times they see ahead.

Is such technological pessimism warranted? The answer to this question has already been answered. Technological developments have already been achieved that will enable the world to move past the 'oil age'. These developments are not so dramatic. They have been around for a long time. At present, they relate primarily to the use of natural gas. But other developments are happening constantly that are not 'high profile'. These include the day-to-day improvements being made in (oil) fuel efficiency in industrial processes, as well as motor vehicle engines, right through to better techniques for the recycling of plastic products. Technological change is not always headline-grabbing. Small changes can have big impacts. There is also little doubt that the process of innovation will continue and that sometimes the outcomes will attract a headline. We can't expect to know what they will be ahead of time. But we can be sure that the price mechanism and reliance on people seeking out opportunities to economise, substitute and innovate is the best way to ensure that the process continues.

There is also no doubt that some promotion of the 'peak oil' 'little green lie' is founded on the pursuit of vested interests. The policies introduced as a consequence of the successful politicisation of 'peak oil' grant advantage to some sectors of society while being harmful overall. For instance, if a person can gain access to a quota of fuel supplied at a low price, then using the 'peak oil' concept to justify oil rationing will be advantageous to them. Similarly, if subsidies are made available for alternative energy research and development or production, then those who are engaged in those endeavours will be made better off. It would be in the interests of these groups so advantaged to encourage the acceptance of the 'peak oil' concept. This will be an especially effective strategy politically if the gains from policies instigated as a result can be concentrated in small politically powerful groups within society at a cost that is spread thinly across the whole of society. For instance, promotion of alternative energy sources will confer gains on the relatively small group in society who manufacture wind turbines and solar panels but spread the costs across all energy consumers. The argument that subsidies to renewable energy sources are simply counteracting assistance provided to the politically powerful oil industry can be dismissed as a case of 'two wrongs don't make a right'. The existence of subsidies to the oil industry is rather an argument for their removal.

Whatever the motivation, the key conclusion to be drawn about 'peak oil' is: 'Don't panic'. First, even under conventional Reserves-to-Production (R/P) ratio logic, it is clear that we haven't reached the point of 'peak oil' production. Second, if one extends the definition of reserves to include fossil fuel substitutes for oil, we are even further from the time of 'peak oil'. That gives the world a lot of flexibility in terms of the time available to develop, prove and introduce alternatives to provide for the needs that are currently being met by oil. In a saying that is well known in the oil industry, the chances are that just as the Stone Age didn't end because we ran out of stones, so too the oil age will not end because we run out of oil.

2: Renewable Energy

Proposition: Renewable energy production should be stimulated.

Non-renewable energy supplies are being depleted so quickly that we will soon experience power shortages. Non-renewables are also 'dirty' sources of energy. Renewable energy production must be stimulated to ensure the on-going supply of clean energy.

BUT

Renewable energy sources are limited in their short to medium term potential to meet demand. Picking the 'winners' to be stimulated is likely to be mistaken with rapidly evolving technological change. Renewables have their own environmental downsides.

Change is constant

With so much discussion about renewable energy of late, one would be excused for thinking that it was a new concept. Biofuels, wind, solar, geothermal, tidal, etc. are widely promulgated as the solutions to what we are told are our energy woes.

In fact, renewable energy sources have been integral to human existence for as long as humanity has existed. The food that maintains our species is grown with the energy input of the sun converted via photosynthesis. Fires fuelled by wood and animal dung have been used for millennia to cook our food and keep us warm through cold winters' nights. Draft animals, fuelled by pastures, have been used until recently (relative to the age of humanity) as the primary source of power in our agricultural and land transportation industries. Where conditions proved favourable, grains have been milled using the power of wind and water. From this perspective, the non-renewable, fossil fuel energy sources are the aberration. But that aberration has had profound consequences for humanity. Fossil fuels – first coal and then oil – were key to the Industrial Revolution and the boom in trade that facilitated so much of the last 200-year period of intensive economic growth and the associated improvements in standards of living for so many people. Oil, in particular, provides a source of energy that is exceptionally compact (in terms of its power to weight/volume ratio) and flexible in its use.

Over time, therefore, the mix of energy sources has varied dramatically. That mix continues to evolve through time. So too does the mixture of energy sources vary across space and uses. Particular circumstances, including the scale of application and relative ease of access, mean that different energy sources will have the comparative advantage. Hydro power is competitive in water-rich countries with steep mountains. It can be conveniently switched on and off to provide power in 'peak-load' conditions. Coal-fired thermal power is cheaper when generated close to coal mines and used to supply the 'base-load' of an electricity grid. Wind, solar, tidal and geothermal power are dependent on natural conditions. Scale of use may also be an important factor. For instance, gas may be preferred over coal for the firing of a small metal fabrication foundry.

So, over time, space and application, the sources of energy have varied and continue to vary. A key part in determining the mix is the relative cost of the different sources. Hence, in the early 1980s after the peaking of the world oil price and the threat of trade embargoes, electricity generation facilities around the globe engaged in a shift away from oil-fired plants to coal-fired plants. The relative costs switched and as a result, so did the relative mix. Technological advance has a big role to play in the determination of cost relativities. The development of improved methods for injecting pulverised coal into the boilers of electricity generation units in the last two decades has meant that the 'productivity' of coal as an energy source was improved, thereby effectively lowering its cost per unity of energy output and improving its competitive position relative to the gas- and nuclear-fired alternatives.

Into this set of factors affecting the efficient production of energy from many different sources can be added the influence of specific government policies. For instance, decisions to commit to the nuclear power option have been in part determined by governments being willing to underwrite the insurance risks facing such plants. Conversely, the decision to close German nuclear power plants in the wake of the Fukushima accident in Japan is also a result of government policy.

The addition of government policy as a causal factor has been somewhat motivated by the 'little green lie' considered in the previous chapter: that the world is running out of time to adjust to a situation where there is no more oil left. The policy initiative is thus designed to encourage the development and introduction of renewable energy sources into the mix so that we can rest assured that the world has an energy future after oil and coal.

The policy imperative is also motivated by concerns about the environmental consequences of non-renewable energy sources. Fossil fuels, in particular, are characterised as environmentally harmful by those seeking to motivate policies to promote renewables. This is both in the local and regional context (smog, sulphur dioxide and nitrous oxide contamination of the atmosphere) as well as globally (as a prime factor in causing anthropogenic climate change).

Hence the proposition put forward by the renewable energy 'little green lie' is that by promoting renewable forms of energy, the world will avoid the energy precipice we face as the oil wells run dry and we will also avoid the environmental catastrophe that non-renewables are causing. An adjunct to this 'little green lie' is that the world's energy demands will be capable of supply without recourse to fossil fuels if renewable energy sources are promoted in this way.

In response to the renewable energy proposition, this chapter will ask if renewables have these twin capacities of supplying the world's energy needs while avoiding environmental harm.

Renewables in the mix

In the global energy picture, renewables are notable for their relatively minor contribution to the total supply. In 1973, the proportion of the total primary energy supply provided by the sum of the renewable energy source categories (geothermal, solar, wind, combustible renewables and waste, and hydro) was 12.5 per cent.[1] By 2009, that proportion had risen to 13.3 per cent. So even though the overall amount of energy supplied from renewable sources had grown over that 36-year period (total supply was 6115Mtoe[2] in 1973 and had doubled to 12,150Mtoe by 2009) the percentage contribution had increased only marginally. The world's reliance on fossil fuels fell from 86.6 per cent to 80.9 per cent, with an increase in nuclear power generation making up the difference.[3] It is noteworthy that within the fossil fuel contribution, different forms altered their relative significance. While in 1973 oil contributed 46 per cent to the total that proportion fell to 33 per cent by 2009. In contrast, coal's proportion rose from 25 to 27 per cent and gas went from 16 to 21 per cent.

In particular it is worth noting that, within the global contribution mix, the geothermal, solar and wind category made up 0.1 per cent in 1973 and had grown to 0.8 per cent by 2008. This amounted to an 800 per cent increase over 36 years. But in the overall scheme of things, their contribution has been very small. Put simply, for renewables to

---

1 International Energy Agency (2011), 'Key World Energy Statistics', http://www.iea.org/textbase/nppdf/free/2011/key_world_energy_stats.pdf
2 Millions of tonnes of oil equivalent.
3 Nuclear power's contribution to total world energy supply rose from 0.9 per cent of total supply in 1973 to 5.8 per cent in 2009.

make more of an impact in overall supply, the conditions that drive the choice of energy source will have to be very different to those currently prevailing.

In order to justify governments stepping in to alter prevailing conditions, the first thing that has to be established is what has gone so terribly wrong with the process by which the energy mix is currently determined?

What has gone so wrong?

Choices between alternative energy sources – be they renewable, non-renewable fossil fuels or nuclear – are resource allocation choices of the type made by people all around the world, every day. For example, as individuals, we choose between beef and chicken to eat, cotton or nylon to wear, and to work as secretaries rather than shop assistants. And corporations choose between locating in one city over another, automating their production line or hiring more people and marketing their products via the Internet or in a chain of retail outlets. The essence of these choices is the interaction between people buying and selling all sorts of combinations of resources in markets.

When we make a decision in a food market to buy beef rather than chicken for the evening meal, we consider how much we expect to enjoy each type of meal and compare that against the amount we'll have to pay for each. Our goal is to choose the option that gives us the most (net) satisfaction.

For the seller, they will need to work out the costs they have had to pay for all the resources they have used to bring the good to market and then offer their product for sale at a price that compensates them for those costs. They will only choose to produce and sell their goods and services if they can make a profit from doing so.

Normally, when an exchange between buyer and seller takes place, we can be confident that both parties consider themselves to be better off as a result. They wouldn't be part of the exchange if that wasn't the case – presuming, of course, that the exchange is voluntary. We can also be satisfied that through the exchange, resources have been transferred to relatively higher value uses. This is because the buyers have signalled that the value they hold for the resources embodied in the good they are buying (their willingness to pay) is greater than the value of those resources in their next best uses. This 'next best use value' is what sellers have had to pay for the resources they have used in producing the good: the sellers have had to induce the owners of those resources to part with them and so have had to pay in costs at least as much as the previous resource owners have valued them.

In this exchange process, because the buyers are willing to pay at least as much as sellers are willing to accept, a mutually beneficial exchange occurs that ensures resources move to higher valued uses.

By entering the market, buyers and sellers provide information to everyone about the value they have for resources. Markets therefore act to transmit these relative value signals at very low cost. In this way, the costs of searching, negotiating and exchanging – the so-called 'transaction costs' – are minimised through voluntary action in markets. With lower transaction costs, more trades are worthwhile and hence more mutual benefit is enjoyed by buyers and sellers from the exchange.

The process of market exchange has been so successful as a means of coordinating the allocation of resources within societies that more and more of them have adopted it. The fall of the Berlin Wall and the rise of the market-based economy in the People's Republic of China have been world-changing examples of the recognition of the strength of voluntary exchange of private property rights. Much of the growth in wealth enjoyed by people around the globe over the past 200 years can be attributed to the gains enjoyed from market trade.

Yet a similar world-wide trend has been for governments to intervene in the operation of some markets. Concerns that some markets are 'failing' to deliver outcomes that are in society's best interest are manifested in policies that see the privately motivated actions of buyers and sellers modified by the coercive powers of government.

These concerns are focused on two key aspects of resource allocation. The first is that the amount of wealth being generated for people by market processes could be improved. This is an 'efficiency' argument. The second is that the distribution of wealth that results from market processes is not to the satisfaction of society. This is an 'equity' argument.

And in energy markets ...

These concerns about markets in general are frequently expressed in the context of energy markets. Two primary points are used to support government intervention.

The first concern is that the supply of non-renewable energy sources, especially crude oil, is running out 'too quickly'. This is the 'peak oil' worry that the world will be left without sufficient energy sources to satisfy its needs in the future. Within this concern, there are both efficiency and equity aspects. In efficiency terms, the argument is that society could be better off if the rate of use of non-renewables was slowed down and the rate of renewable development and implementation was speeded up. In equity terms, the concern is that future generations will be worse off because the current generation is using up all the non-renewable energy resources.

The second concern underpinning arguments to support government intervention is that non-renewable energy sources are 'too dirty'. Again there are efficiency and equity dimensions involved. From the efficiency perspective, the pollution associated with non-renewables makes people worse off to the point where, if it were reduced, there would be overall gains in well-being (non-monetary wealth) to be enjoyed. The equity angle again relates to future generations being disadvantaged relative to the current generation.

Policies responding to these points come in a variety of shapes and sizes. Regardless of their characteristics, all interventions by governments mean that the relative value-signalling processes inherent in market exchanges are compromised. The information transmitted by purchases via the prices paid and received is overruled and alternative signals are established. Essentially this means that the information provided (voluntarily) by individuals engaged in trade is displaced by information that is generated by the state. What does the collective 'wisdom' of the state know about energy sources that self-motivated individual actors don't know?

The 'too quickly' concern arises because, it is argued, the self-interest of people now ignores the virtues of resource conservation that can be seen from the society-wide perspective taken by the state. The 'too dirty' concern also involves individuals acting to better themselves, ignoring the wider consequences of their choices.

The question that has to be raised with any government intervention that superimposes its resource allocation signals over the top of private market signals is whether or not the actions taken actually improve the well-being of society. It is not sufficient simply to *assert* the principles of market 'failure' to achieve society's goals, it is also necessary to *demonstrate* that an intervention does not make society worse off: That is, the proposed intervention must be shown to deliver benefits to society that are greater than the costs it imposes.

The real prospect is that the information governments work with to devise its interventions is even more problematic than that generated in private markets. In addition, the incentive structures that operate within governments may be perverse to the achievement of improved overall social well-being.

Government action
The response to concerns about the energy mix by governments around the world has involved a range of policy instruments. Regulations have been formulated to restrict or even prevent some non-renewable energy uses, while others have mandated renewables' use. Payments of subsidies have been made to encourage the production

and consumption of renewable energy. In other jurisdictions, additional taxes have been levied on the production or consumption of non-renewable energy.

The use of ethanol and bio-diesel in the liquid transportation fuel market has been mandated in some countries. For instance, in Brazil, the proportion of ethanol mixed with petrol at the service station is set at 25 per cent.[4] Tax credits are provided by the US federal government to companies that blend petrol with ethanol. The cost between 2005 and 2009 was USD17 billion. In 2010, the single-year cost was USD5.4 billion.[5] In Australia, the federal government will require electricity utilities either to generate 20 per cent of their total output from renewable sources by the year 2020 or to hold renewable energy certificates (RECs) to show that they have offset any power generated from non-renewables beyond 80 per cent.

All of these and similar measures elsewhere have the impact of signalling that the costs of renewable energy are lower, relative to non-renewables, than the market would otherwise indicate. The potential consequences are significant. There is a substitution away from non-renewable energy sources towards renewables. Although the government may use policy measures to signal that renewables are relatively cheaper than non-renewables, the actual market costs of energy are unchanged. That means the overall cost of energy rises as a result of the policy interventions to change the supply mix. The higher cost is either borne by consumers of the energy through higher prices, by taxpayers who fund the payment of subsidies or by some combination of the two.

The extent of the cost differences between renewable and non-renewable energy sources are worth considering. CSIRO[6] data indicate

---

4 http://www.economist.com/node/16952914
5 http://www.economist.com/node/16492491
6 Graham, P. (2006), 'The heat is on: The future of energy in Australia', CSIRO, Canberra. (http://www.csiro.au/Organisation-Structure/Flagships/Energy-Transformed-Flagship/The-heat-is-on-the-future-of-energy-in-Australia---PART-ONE.aspx).

a cost per MWh[7] for electricity generated in a black coal thermal power plant in Australia of around AUD28-38. That sets the benchmark for costs against which the renewables can be compared. Hydro power generation is competitive, depending exactly on the rainfall and terrain characteristics applicable at any given site. For some sites, hydro is cheaper than black coal, but the average cost is AUD55 per MWh. Both wind and solar electricity generation, are higher cost than black coal. Wind is in the order of AUD63 per MWh, while solar thermal is nearly AUD85 per MWh and photovoltaic power costs in the order of AUD120 per MWh.

What are costs?
Costs are paid by producers of power to induce the owners of resources used in the production of power to give up their other uses of those resources. So the costs paid are a reflection of the value to society of what is being given up in order for the power to be produced. If a resource is particularly costly, then that is a reflection of the value it provides in its next best alternative use. If it is a particularly scarce resource with high-value alternative uses, the amount that will have to be paid by the new user to compensate its current owner for giving up those high-value alternative uses will be higher. The cost of that power generation will therefore be high as a result.

This provides some insights into the meaning of the difference between the costs of coal-powered electricity and photovoltaic-generated power. The reason why photovoltaic power is so much more costly is because it uses more scarce resources than coal-fired power. These costs are not just related to the source of the energy. For photovoltaic power, that is free and provided by the sun. For coal, the energy source has to be mined, processed and transported, all of which use scarce resources that had other alternative uses that had to be compensated. Many other

---

7   Mega Watt hour.

resources are also involved in producing power from both sources. Land, labour (skilled and unskilled), machinery, and many other types of natural resources are used in differing combinations. The overall costs reflect the relative scarcity of all the resources involved, not just the energy source. The relatively high cost of photovoltaic power indicates that it uses many other resources which, when considered together as a package, are far scarcer than the black coal energy source. An example of the issue of relative scarcity has emerged recently in the role of the so-called 'rare earth' minerals. These resources are important in (amongst other uses) the production of electronic componentry and batteries, two critical components in the renewable energy story. Prices of the rare earths have recently increased significantly with growing demand. In addition, export restrictions imposed by the Chinese government have also had an upward impact on prices. This is because Chinese supplies of rare earths dominate the current market.

Put simply, policies that encourage the use of more expensive renewable sources of energy increase the rate of use of other resources that are scarcer than the non-renewable being replaced. The policies therefore shift the pattern of resource use toward those which are scarcer.

Ripple effects

Policies to promote renewables have impacts beyond the energy sector of an economy. Their impacts spread like ripples in a pond when a stone is thrown into the middle.

As explained, the costs of resource use signal the value of a resource in its next best alternative use. By enticing more resources away from their alternative uses, policies that promote renewable energy sources have impacts on the markets for goods and services that were the alternative uses. Nowhere has this been seen more clearly than in food markets following the introduction of policies to encourage the production of ethanol as a liquid fuel substitute for crude oil.

Ethanol is produced in Brazil from sugar cane. The alternative use of the sugar cane resources used to produce ethanol is to produce sugar. With mandates to produce ethanol instead of sugar, the supply of sugar dropped and the price went up in response. Sugar consumers were therefore made worse off, having to pay more for diminished supplies. In the US, ethanol is made from maize (less efficiently than it can be made from sugarcane in Brazil). The US policy of providing tax credits for ethanol production meant that more maize was supplied to produce ethanol and less was available as an animal feed and to make tortillas. Cattle feed and meat prices increased as a result. Furthermore, land that had previously been used to grow other grains such as wheat was planted to the now 'artificially' more profitable maize. International grain markets responded sending not just wheat prices up but also rippling on to rice prices. The consequent 'food crisis' that resulted was only relieved by the pain of the global financial crisis[8] and further peaks in world food prices have been experienced in 2011 as the world economy has recovered. In 2010, the US government introduced the new Renewable Fuel Standard (RFS2) that limits conventional ethanol to 15 billion gallons of the annual 36 billion gallons of renewable fuel that must be used for transport by 2022.[9] The signal given by government regarding the relative scarcity of energy resources proved so confounding that another policy signal has been imposed on top of the first. It provides a good illustration of the complexity of information that is generated within markets and how difficult it is for governments to be able to 'do better than the market'.

Pollution costs

The second element of concern that has stimulated government intervention to stimulate the production of renewable power revolves around the view that non-renewable fuel sources are 'too dirty'. This

---

8   http://www.fao.org/worldfoodsituation/en/
9   http://www.economist.com/node/16492491

concern has been particularly focused on the production of greenhouse gas emissions (predominantly carbon dioxide) when fossil fuels are burnt. There is more on this issue in Chapter 9.

But there are also environmental concerns resulting from the extraction of non-renewable fossil fuels including the impacts of open-cut coal mines and tar sands mines and processing plants on ecosystems, air quality and water quality and the risk of spills from off-shore oil wells causing damage to marine environments including beaches and estuaries. Spills occurring during the transportation of crude oil have also been some of the most prominent environmental disasters of recent decades, with the grounding of the Exxon Valdez in Alaskan waters being the most damaging.

Furthermore, environmental harm does arise from the burning of fossil fuels beyond any impacts on global climate change. Motor vehicle transportation and electricity generation are primary sources of air pollutants including sulphur dioxide, ozone, nitrous oxides and particulate matter. These are damaging to human health, have an impact on ecosystems and wildlife as well as reducing visibility.

The costs imposed on society by these types of pollution are often not reflected in market-based choices. When choices involve the use of resources that nobody owns – that is, there is no property right defined or defended for the resource – then a market will not form to allocate that resource. It thus becomes 'free' for those who can secure its use. So where coal mining causes the salinity of adjacent ground or surface water to rise so that downstream users are adversely affected or the levels of dust particulates to increase so that the health of adjacent residents is harmed, there may not be any market transactions to ensure these costs are felt by the owner of the coal mine. In other words, the coal mine operator may not have to buy access to the water or the air because of a lack of clearly defined rights to those resources. Similarly, a person driving his or her car to work each day does not bear the costs of respiratory illness in the community caused by air pollution. Nor does the factory owner who powers its operation using electricity generated through the burning of coal bear any costs associated with the poor visibility resulting from the power station's emissions.

Government interventions are frequent in such circumstances. They primarily result in additional costs being paid by non-renewable producers so that environmental impacts are taken into account. For instance, regulations may stipulate the extent of dust particulates allowed in the air around a mine. To comply with such a regulation, coal mine owners will invest in dust suppression measures that add to the cost of their operations. Governments may also act to define and enforce property rights over environmental assets. For instance, liability laws regarding oil spills mean that a breach would mean the payment of compensation by oil producers and shipping companies to affected parties. This provides an incentive to invest in double-hulled oil tankers and the instigation of low-risk operational procedures – all of which add to the production costs of the non-renewable resource.

Interventions are also in place to deal with the environmental consequences of non-renewable energy use. Cars are required to conform to emission standards. Power stations are regulated in terms of the amount of particulates and various other polluting gases they can release into the atmosphere. These actions impose additional costs on users of non-renewables, passed on from producers through the higher prices of energy that are generated from those non-renewables.

The result of these targeted environmental policies is that the cost of non-renewables is increased to reflect the consequences which their mining, transportation and use have on the environment. In this way, the competitive position of the non-renewables in relation to renewable alternatives is eroded. The choice between renewables and non-renewables is thus taking into account the differences in environmental impacts.

That is not to say that there are policies in place at every non-renewable energy source location and everywhere that non-renewables are used around the world – far from it. In developing countries, in particular, environmental regulations relating to the extraction, transportation and use of fossil fuels are often either absent or are not enforced. This means that non-renewables compete in those settings without all the costs they impose on people being paid for by their users. It will be argued later in this chapter that the failure of governments to intervene to ensure the payment of costs associated with use of environmental resources is more of an argument in favour of getting those interventions right than an argument for encouraging renewable energy supplies.

'Clean' at what cost?

Renewable energy sources have some undoubted environmental advantages over non-renewable fossil fuels. Solar and wind power have the potential to generate electricity without the on-going emission of ozone, sulphur dioxide, particulates, etc. But two important caveats need to be kept in mind.

The first is that renewables should not be seen as 'innocent' in the environmental pollution stakes. For a start, even renewable power sources are responsible for greenhouse gas emissions; albeit at a level lower than fossil fuels. This arises because renewables involve capital construction that involves the emission of greenhouse gases. For instance, solar photovoltaic power emits 76 grams of carbon equivalent per kWh.[10] This compares to black coal-generated thermal electricity with produces 357 grams and natural gas at 188 grams. Wind power does better at 13 grams and hydro (in Germany) is rated at only 6.3 grams. The production of ethanol from maize offers only minor reductions in greenhouse gas emissions compared with fossil fuels

---

10 Spadaro, J., L. Langlois and B. Hamilton (2000), 'Greenhouse Gas Emissions of Electricity Generation Chains: Assessing the difference', *IAEA Bulletin*, **42** (2): 19-24. (http://www.iaea.org/Publications/Magazines/Bulletin/Bull422/article4.pdf)

because the cultivation and fertilization processes involved in growing maize are heavy fossil fuel users in themselves.

Renewable energy sources have other environmental costs. Some of these may not be paid for by their operators in the same way as fossil fuel users may escape paying some of the costs they impose on the environment. There are environmental impacts from the mining and processing of the 'rare earth' minerals[11] important to the electronic componentry and batteries that back up renewable supplies. There is visual pollution associated with large-scale wind and solar farms. Wind turbines are criticised for disrupting bird flight paths and creating a noise nuisance for local residents. Hydro power development has ceased in most developed countries, because of the environmental damage done to waterways when they are regulated and the loss of valley floors when they are inundated.

The environmental implications of biofuel production are also worth noting. The payment of subsidies for ethanol production has, as indicated earlier, given an impetus to sugar and maize growing. Their areas under cultivation have expanded and these additional acreages have not just come from competing crops. Additional plantings have been made in areas that were otherwise set aside as natural vegetation. For instance, in the USA, prairie lands, set aside under the Conservation Reserve Program by which farmers are paid not to cultivate, were brought into maize production. The payments made under the Conservation Reserve Program were insufficient to compete against the boosted returns from maize growing. Hence the encouragement of ethanol production inadvertently led to the loss of nature protection areas and, potentially, endangered species habitat.

The second caveat that comes with the environmental credentials of renewable energy goes back to the point made earlier about the higher

---

11 The environmental damages caused by the mining and processing of rare earths are cited by the Chinese authorities as reasons for restricting their production and export.

costs of production associated with renewables. Yes, renewables may come with some environmental advantages but they do so at a cost. That cost is the forgone opportunities of resources used in renewable production that could have been used in other valuable uses elsewhere in society. Are the environmental advantages worth those costs? When the cost difference between black coal-generated electricity and photovoltaic-generated power is in the order of four times (AUD30 *vs* AUD120), it is a very serious question to be answered.

Too slow?

So to recap the two dimensions of the renewable energy 'little green lie': Renewable energy sources need to be encouraged through government intervention because the process of shifting from non-renewables to renewables is 'too slow' and because non-renewables are 'too dirty'.

As was detailed in Chapter 1, price plays a critical role in determining the speed at which different resources are used and are substituted one for another. An increasing price indicates increasing relative scarcity and is a signal that the rate of use of that resource should slow down and substitute resources start to be in increasing use.

For non-renewable resources, the price paid now also reflects the scarcity of the resource in the future as well as now. The reason for this goes back to the incentives facing those who own a non-renewable resource. They have, fundamentally, two options about how to use their resource: Extract it and sell it now or hold it in the ground to be sold at a future time. The values enjoyed under these two options are called the 'use value' and the 'asset value'. Resource owners therefore have to think carefully about the likely prospects of price changes into the future in order to evaluate the asset value and how it rates against the use value. Because the asset value captures the prices expected to be paid by future users, current owners of the resource take into account the interests of future users. If it is expected that the resource will be very scarce in the future, current owners will hold off extracting their resource now in order to take advantage of the higher asset value. The implications of that choice are worth noting. With more resource owners holding off extraction, current supplies are reduced and current prices go up. And with more being held back for future delivery, expected future prices will be lower. In this way, the competing demands of current and future users are reflected in the choices of current resource owners. The impact is that the asset value of a non-renewable resource acts to protect the interests of future users in today's markets.[12]

Hence, with increasing scarcity of a non-renewable resource, the price rises in a 'turbo-boosted' way, reflecting not just current scarcity but also the anticipated increasing scarcity to be faced in the future. The signal this sends to buyers is to shift their demand to substitutes (the next lowest cost renewable or non-renewable resource) and for sellers to compete in the development and supply of substitutes (both renewable and non-renewable) at lower cost. Therein is the process of continual adaptive adjustment required to deal with changes in relative scarcity. The actions of millions of buyers and sellers motivated by making themselves better off send signals through society about the continuously changing picture of relative scarcity.

Interventions that mandate change or seek to manipulate the process of change are likely to be counterproductive. This is because, first, governments are highly unlikely to have the amount and detail of information necessary to manipulate successfully the complex process of change that is constantly occurring in markets. The almost instantaneous feedback loops apparent in market information-generation processes are unavailable to regulating bureaucracies that work with long lead and lag times between the collection of input data and the analysis of consequences from policy initiatives. The incentives for governments

---

12 This is the process of adjustment over time first outlined by Harold Hotelling and introduced in Chapter 1 of this volume.

and their bureaucracies to 'get it right' for society are also weak. The danger is that interventions will be used to further political and personal interests rather than the well-being of society as a whole.

Interventions can therefore be seen more as interruptions and distortions to the process of change that is constantly emerging in market transactions. This is disruptive to the signalling of resource scarcity that prices are continually providing. Decisions are therefore made on the basis of prices that are sometimes distorted by government intervention-driven information rather than the aggregation of the wisdom of all those involved in the market, and who therefore have an incentive to 'get it right' for themselves and thereby for society. These decisions will be sub-optimal and thus costly. For instance, subsidising solar photovoltaic arrays on suburban rooftops is costly because resources that could have generated higher value elsewhere in the economy were used.

An all-too-often-witnessed example of governments falling into the trap of intervention comes in the form of 'picking winners': that is, choosing the technological option or industry that is most likely to deliver specified policy outcomes. In the renewables area, this usually involves the selection of which renewable technology to support through the payment of subsidies. Given the large difference between the costs of wind power generation and solar photovoltaics, it is hard to understand why an intervention to support renewables would invest in photovoltaics beyond that needed to promote research. Yet schemes around the world are in place demonstrating that governments believe that these are potential 'winners'. Why would that be the case? Perhaps it is because there is an effective political lobby group supporting the solar panel industry? Perhaps because it is an investment that anyone with a rooftop can make and feel that they have contributed to overcoming the world's energy 'crisis'? This satisfaction may be reason to vote for the government that instigated the subsidy. But it is a very costly choice.[13]

Irrespective, the lesson is clear. Market forces provide the signals necessary to see the substitution of renewables for non-renewables over time. The interventions by government to 'force' change will impose costs upon society. These costs are representations of the resource uses that could have been enjoyed had the interventions not taken place – more schools, hospitals, shoes, cars, etc.

Too dirty

The environmental consequences of power generation need attention, frequently through the collective action that can be applied by government. The mining operations extracting fossil fuels, rare earths, etc. have environmental impacts that should be managed in ways to make sure society is not made worse off. Similarly, the same applies for the transportation and consumption of these fuel sources. There is an array of environmental management techniques that is available for the tasks. Cap-and-trade schemes for pollutants and offset schemes for biodiversity impacts are examples of trading solutions to these challenges. Regulatory solutions are also available and are most commonly used in developed countries.

Encouraging renewable energy production as a means of dealing with the environmental impacts of non-renewable energy sources is a very 'blunt instrument' indeed. It is far removed from the policy target – for example, reducing the dust pollution caused by a coal mine – and only indirectly related. It is thus very unlikely to have the desired policy outcome and is also likely to bring with it distortions to the energy market (as indicated above) that make it counterproductive.

---

13 The bankruptcy in 2011 of Solyndra LLC, a US solar panel producer that had received a $528m federal government loan, is indicative of the problematic nature of politically driven subsidies for renewable energy ventures that are not commercially competitive (see: http://www.washingtonpost.com/politics/solyndra-saga-and-its-key-players/2011/09/30/gIQAZZDYAL_story.html).

The Tinbergen principle[14] pertains: Select a targeted policy instrument for each separate policy goal. Trying to use one instrument to achieve multiple goals is likely to be counterproductive.

A 'little green lie'?

Interventions to hasten the switch from non-renewable sources of energy to renewables are likely to be costly to society and run the risk of distorting the signalling processes that are active in markets regarding the relative scarcities of resources. The smooth transition from different mixes of energy sources that is afforded by the constant adjustments to these relative scarcity signals is thus put in jeopardy. For instance, the potential for natural gas to fill a 'gap' between oil and a renewable energy source is not factored into the interventions encouraging renewables. Furthermore, the 'picking of winners' that the government will support is not only likely to be misinformed and driven by political interests, but it is also likely to distract from the development of options that are still 'ideas' in the minds of creative engineers and scientists.

The motivations for supporting this 'little green lie' relate to the fears held by people for a time when energy ceases to be available and when they witness the environmental consequences of some non-renewable energy sourcing, moving and using activities.

A key message from this and the previous 'peak oil' chapter is that the energy starvation fear can be readily dismissed. More likely are short-term disruptions to energy supplies caused by natural disasters and global political instability. Such disruptions are handled more cost-effectively through the development of storage facilities both by the private sector and national authorities.

The consideration of environmental impacts of energy given in this chapter delivers the conclusion that environmental management should

---

14 Tinbergen, I. (1950), *On the theory of economic policy*. Elsevier, North Holland.

involve the careful design of policies that are directly targeted at the environmental issue of concern. Using assistance for renewable energy sources to deal with the environmental consequences of non-renewables is likely to be counterproductive to society as a whole.

The renewable energy 'little green lie' can be readily dismissed on both the 'too fast' and the 'too dirty' 'fears'.

Certainly there are political pressures to explain the telling of this 'little green lie'. Industry support, be it for renewable energy generation or motor vehicle manufacturing or agriculture, can usually be associated with well organised and politically influential lobby groups. The renewable energy (green) lobby, particularly in Europe, and there particularly in Germany, has been very successful in securing subsidies to assist their otherwise uncompetitive businesses. Dispelling that source of the 'little green lie' is a politically vexed challenge.

3: Index-based Consumption Decisions

Proposition: Consumption choices need to be informed by products' environmental characteristics such as their 'food miles', 'ecological footprint', 'embodied energy', 'virtual water' and 'carbon footprint'.

People need to be aware of the impacts they have on energy, the ecology, water, climate, etc. when they buy goods and services so that they can reduce their impact on those resource. Each resource is scarce and valuable. We need to conserve them, especially for future generations.

BUT

By focusing on just one scarce resource (water, energy, etc.) in their consumption decisions, people can ignore their impacts on other scarce resources. A 'false economy' results. When impacts on multiple resources are combined in an index, more distortions arise. Prices do a better job of signalling resource scarcity.

Scarce resources

In the previous two chapters, the focus has been on 'little green lies' that are founded on the fundamental premise that non-renewable energy resources (notably crude oil in Chapter 1) are worryingly scarce. So much so that action needs to be taken at a government policy level to make sure we are not left 'high and dry' as reserves of non-renewables are run down. The other imperative for policies that support the accelerated replacement of non-renewables with renewables is the comparative environmental pollution profiles of the two energy source types. There is a belief that 'clean and green' deserves support when compared with the actual and potential environmental harms done by non-renewables.

Social concern for resource scarcity has not been limited to energy sources. Different resources in different places and at different times come in to and out of focus as being worth worrying about because different groups within society achieve prominence at different times. One way in which these concerns have become evident is through the growth of the preparation and use of indices that indicate the extent to which a specific resource is being used when a good is consumed. The object of this process is to make consumers aware of the impact they are having on the resource of concern so that they can change their choices in recognition of the additional information carried by the index.

For example, in times of drought and in areas where water scarcity is a pressing constraint on people's well-being, the impact on the water resource of consumption choices is top of mind for many. Hence the conception of the 'virtual water' index. The Virtual Water website[1] defines it as:

> ... the amount of water that is embedded in food or other products needed for its production.

We are thus able to know that drinking a cup of coffee involves the use of 130 litres of water compared to a cup of tea's 'water footprint'[2] of 27 litres. On a grander scale, the production of a tonne of cereals (on average across the world) involves the use of 1,644,000 litres of water compared with vegetables which require 322,000 litres per tonne. Animal products are more water-intensive in their production. For

---

1   www.virtual-water.org
2   The term 'water footprint' is used to indicate the amount of virtual water used in the production of any specific good or can be extended to mean the aggregate amount of virtual water used by specific regions or nations in their production processes.

instance, production of a tonne of cheese is claimed to use 5,000,000 litres of water and a tonne of beef, 16,000,000 litres. The list of products with a 'water footprint' is not limited to foods: A 32 megabyte computer chip weighing 2 grams has a virtual water use of 32 kg (that is, 32 litres).[3] Under the virtual water methodology, water use is divided into green, blue and grey categories. Green is water sourced from rainfall, blue is from surface and groundwater sources, and grey is the amount of water required to assimilate any pollutants associated with the good's production so that environmental standards are met. So in the case of cereals, as an example, 75 per cent of the virtual water requirement is met by rainfall while 14 per cent is 'blue water' with the remaining 11 per cent being 'grey' water.[4]

Virtual water 'calculators' on websites[5] allow individuals to estimate their individual water footprints by answering questions about their dietary habits ('how many cups of coffee do you have a day?'), their personal hygiene routines ('how many showers do you take a day?'), yard requirements ('what is the capacity of your swimming pool?') and a general catch-all for other products' water usages ('what is your gross yearly income?').

While the thrust of the peak oil (Chapter 1) and renewable energy (Chapter 2) 'little green lies' is toward stimulating government action, the consumption index phenomenon is directed toward individual consumers. There is an element in the use of indices that is aimed at 'informing' governments in their deliberations about resource use policy. However, the emphasis is more on convincing individual consumers to make changes to their consumption and lifestyle choices.

---

3   Hoekstra, A.Y. (2003), 'Virtual water: An introduction' in A.Y. Hoekstra (ed.), *Virtual water trade: Proceedings of the International Expert meeting on Virtual Trade*, Value of Water Research Report Series No 12, IHE Delft: http://www.waterfootprint.org/Reports/Report12.pdf
4   'Grey' water from cereal production relates to the pollution caused by the use of fertilizers, herbicides and pesticides.
5   http://www.virtual-water.org/index.php?option=com_wrapper&Itemid=8

This trend is apparent in a wide range of specific resource 'indices'. Energy is a prominent case, with the emergence of concepts such a 'embodied energy'. This estimates the amount of energy used in the production of goods and services. It has gained most traction in the building industry, with different building materials having differing levels of 'embodied energy'. For example, Urban Ecology Australia Inc. argues that:

> A building's embodied energy – the energy expended to create it, and later remove it – can be minimised by constructing it from locally available, natural materials that are both durable and recyclable, and by designing it to be easy to dismantle, with components easy to recover and reuse or recycle.[6]

As with 'virtual water', there are calculators available to estimate the embodied energy in different types of buildings.[7] The energy inputs include construction and demolition activities to give a 'life-cycle'[8] appreciation of the energy requirements involved in any particular dwelling. The results are in the form of British Thermal Units (BTU) per building. These can also be presented as BTUs or other units of energy, such as kilojoules per unit of space to provide a scale standardisation factor. For instance, for a typical single family residential house, the embodied energy is around 700 MBTU[9] per square foot. Clustering houses into attached groups comprised of two to four residences reduced this to 630 MBTU per square foot.

---

6   http://www.urbanecology.org.au/topics/buildingsembodiedenergy.html
7   http://www.thegreenestbuilding.org/
8   Life Cycle Analysis is inherent in both the virtual water and the embodied energy concepts. Life Cycle Analysis is 'a recognised instrument to assess the ecological burdens and impacts connected with products and systems, more generally with human activities' according to the *International Journal of Life Cycle Analysis*. It involves tracking through the resource use history of production and consumption activities cataloguing in a 'cradle to grave' fashion the direct and indirect uses of the resources involved.
9   One thousand (M) British Thermal Units (BTU).

The embodied energy index has been extended to cater for interest in the carbon (and hence climate change) impacts of buildings. A consortium of consultants, researchers and lawyers[10] has developed a calculator that first estimates embodied energy and then converts that into carbon emissions. The calculator was initially designed as a tool to be used in the implementation of policies designed to impose a tax on greenhouse gas emissions.

Carbon foot-printing has developed independently of the embedded energy concept. Rather than focusing on the extent to which a product uses a scarce and hence valuable resource, the carbon footprint indices estimate the extent of greenhouse gas emission caused during the production and consumption (or use) of a good. The scarce resource being considered is the capacity of the world's atmosphere to take up greenhouse gas emissions without causing climate changes. Carbon footprint calculators abound[11] for activities ranging from air travel to household waste disposal. There are even calculators designed for children to see the impact of taking a shower instead of a bath or watching half an hour less television each day.[12]

Broader indices

A problem that is very quickly evident with the development of single resource indices is the potential for them to work against each other. For instance, if a consumer is concerned about water scarcity but is also worried about the energy future of the world, he or she may want to make decisions on the basis of both the virtual water *and* the embodied energy characteristics of a product. It may be that in the comparison of two competing products, one provides better scores on both indices. The problem emerges when neither product is able to

---

10 http://www.carboncostcalc.com/asp/index.asp
11 For example: http://www.carbonneutral.com.au/carbon-calculator.html?gclid=COL 60pGhpqgCFQRqfAodU0RaeQ
12 http://www.cooltheworld.com/kidscarboncalculator.php

claim absolute superiority. For instance, a product may use little water because its manufacturer has invested in an energy-intensive production process. What then?

The way that this problem has been addressed has been through the development of broader indices that envelop multiple resources.

One such index is 'food miles'. This presents different foods (and types of foods) to consumers in terms of the distance the food has been transported from its place of origin to its place of sale. While the principal concern of those promoting the food miles concept would appear to be the energy/global climate change consequences of transportation distance, there are also implications for food freshness and support for local agricultural industries and their workforces. Local products are promoted as being less environmentally harmful, fresher (and therefore healthier) and also supportive of maintaining the social fabric and appearance of the countryside. Originating in the UK, the food miles concept has extended its reach internationally.[13]

Other broader indices are not quantitative and so not measured in miles or tonnes, or BTUs. Rather, they are yes/no switches. Products either pass a sequence of tests or they fail. The 'fair trade'[14] index is an example of this type. Those products permitted to be marketed using the fair trade logo have satisfied the conglomerate of 24 organisations that comprise Fairtrade International to achieve a set of product and pricing standards that broadly constitute a goal of 'sustainability'. While being more heavily focused on the livelihoods of producers in developing countries through premium prices paid by developed world customers, Fairtrade product standards also involve environmental hurdles.

Similarly, 'organic' certification for foods has an environmental as well as a health aspect. In Italy, for instance, BioAgriCert[15] certifies

---

13 http://www.choice.com.au/reviews-and-tests/food-and-health/labelling-and-advertising/sustainability/food-miles.aspx
14 http://www.fairtrade.net/about_us.0.html
15 http://www.bioagricert.org/english/index.php?option=com_frontpage&Itemid=1

sellers wishing to market their products as organic. Similar organisations have arisen in countries around the world to act as certifying agents. Other products have similar arrangements. For instance, timber may be certified as being sourced from sustainably managed forests by the Forest Stewardship Council (FSC) which 'provides a credible link between responsible production and consumption of forest products, enabling consumers and businesses to make purchasing decisions that benefit people and the environment as well as providing ongoing business value'.[16]

Such certification schemes are clearly targeted at influencing consumers' purchasing patterns and have become increasingly popular as product-marketing devices.

However, perhaps the best known of the broader indices is the 'ecological footprint'. This index indicates the area of land and sea that an entity (be it an individual, a business, a city or a nation) requires to satisfy its consumption demands, including provision for the disposal of wastes. It is therefore an index that conveys information about demands across multiple scarce resources (as indicated by an area of land and water) and for an aggregation of consumption goods and services. It is used most prominently as a means of communicating to consumers the impact of their choices at a more aggregate level. It goes beyond the single resource indices such as virtual water, and the single product but multiple resource impacting indices such as are inherent in product standard certification.

The Global Footprint Network defines the ecological footprint as 'how much land and water area a human population requires to produce the resource it consumes and to absorb its wastes, using prevailing technology'.[17] This is then compared against the 'supply created by the biosphere, namely biological capacity (bio-capacity) ... a measure

---

16 http://www.fsc.org/about-fsc.html
17 http://www.footprintnetwork.org/en/index.php/GFN/page/footprint_basics_overview/

of the amount of biologically productive land and sea area available to provide the ecosystem services that humanity consumes'.[18] Calculators are provided by the Global Footprint Network to allow the estimation of the footprint of nations, cities, businesses and individuals with the goal of 'ending the overshoot': that is, reducing the extent of the ecological footprint demand so that it remains within the supply capacity of the earth. This goal is motivated by the calculation that the world's current ecological footprint is 1.5 times the size of the earth, and will grow to 2 times by the 2030s given current rates of consumption and growth. A large component – around 80 per cent of the world's bio-capacity – of the ecological footprint is made up of the 'carbon footprint'. This is the amount of land required to sequester carbon dioxide emissions. It has grown from around 10 per cent of bio-capacity in 1960 and constitutes the majority of the increase in the ecological footprint over the last 50 years.

Across the globe, some countries are classified as 'debtors' while others are 'creditors' in terms of their ecological footprints. For example, Canada, Australia, Brazil and Sweden are creditors while others including France, the United States, China and Bangladesh are debtors.

The ecological footprint and many of the other broader indices of resource scarcity stress the concept of sustainability. The indices all seek to operationalize what has been a very nebulous concept to define. The predominant definition of sustainability is that of the Bruntland Commission Report:[19] It called for the well-being of society not to decline across generations. Others have sought to define sustainability in terms of the 'triple bottom line' that accounts for society's impact on financial, social and environmental conditions. The consideration

---

18 Ewing B., A. Reed, A. Galli, J. Kitzes, and M. Wackernagel. 2010. *Calculation Methodology for the National Footprint Accounts*, 2010 Edition. Oakland: Global Footprint Network.
19 *Report of the World Commission on Environment and Development: Our Common Future*: http://www.un-documents.net/wced-ocf.htm

of multiple objectives in a triple bottom line accounting sense is a similar approach to that of another technique used for assisting decision making where a range of different impacts are apparent: Multi-Criteria Analysis (MCA).[20]

MCA generates a quantitative index that has no specific unit of measurement. It works by the analyst establishing a list of 'criteria' that a decision will impact. For instance, forest management options may be judged by the scarcity of the trees being harvested, the amount of soil erosion harvesting causes and the number of jobs created by logging. Those criteria all use different units: Scarcity may be rated on a 1 to 10 scale, erosion measured by tonnes of soil lost and employment by the number of jobs per annum. The predicted impacts of alternative forest management options on these criteria are then standardised using mathematical means. To add all the different criterion scores together needs a way of incorporating the relative importance of each criterion. This is achieved through the application of 'weights' for each criterion. The determination of those weights may be at the analyst's discretion, or somehow devised to reflect the opinions of others, perhaps the decision maker who called for the analysis.

In this way, MCA presents to decision-makers a score for each choice alternative. It has been used as an information base to make decisions as wide-ranging as investments in natural resource management through to the formulation of building product codes.

For example, life-cycle assessments which involve multiple resource impacts are a type of MCA. A life-cycle assessment aims to track the 'cradle to grave' impacts on the full range of resources caused by the production and consumption of a good. When those impacts are aggregated to form an index of overall impact and that index is then used to make recommendations regarding the relative performance

---

20 Also referred to as Multi Attribute Utility Analysis and Program Budgeting and Marginal Analysis.

of products, the assessment forms a MCA. This has been the case in the development of rankings for building products in the UK, the USA and Australia that are then available for use either for voluntary assessment[21] or in building permit applications which require that certain environmental standards be met.

In the context of building materials, Howard and Kneppers[22] conducted a life-cycle assessment (LCA) and concluded that 'LCA practitioners that conduct multi-impact LCA and reach a conclusion and make recommendations must be doing some form of implicit weighting whether or not they realize they are doing so'. In other words, the process of LCA involves impacts across multiple resources that must somehow be standardised for aggregation into a single performance index. This requires the assignment of weights to the different impacts that reflect their relative importance. As such the process of LCA conforms to that which is characteristic of MCA.

Single resource index contradictions

On the surface, using single resource indices would appear to be a useful means for consumers to gather information about product choice. The problem with this approach is that it leads consumers to focus on the impact on just one of the resources used in the production of a good or service. The example of two different indices working in opposite directions, provided earlier, demonstrates the problem with using just a single index. However, the problem is much broader than this.

First, any index that is based on the physical units of a resource that are used necessarily masks the complexity of resource scarcity. Put simply, the same unit of a resource will be more or less scarce depending

---

21 Compliance is voluntary with the Australian Green Building Council's 'Greenstar' system which considers the environmental impacts that are embodied within materials (see: http://www.gbca.org.au/green-star/green-star-overview/).
22 Howard, N. and B. Kneppers (2011), 'Weighting for LCA based Tools, Methods and Ecolabels – Practical but Contentious!', paper presented at the 7th Australian Conference on Life Cycle Assessment, March.

on a whole range of circumstances. For instance, the litre of water used to grow coffee in the damp tropical forests of Costa Rica has a different level of scarcity to a litre of water being used to brew cups of coffee in Saudi Arabia. Adding together the litres of water used to produce a cup of coffee as a measure of water scarcity is therefore quite meaningless.

Furthermore, the production of any good or service involves the use of a multitude of scarce and valuable resource inputs. Focusing in on one of those resources – or even several resources – takes the risk of ignoring the consequences of use for the stock of other resources. Choosing on the basis of an alternative's carbon footprints for instance, may ignore the consequences of a product's use of inputs other than energy. They may be natural resources (copper, rare earths, water, etc.) or labour and capital resources. Hence the blinkered view of choice provided by a single resource index may well drive undesirable choices, when broadly considering what is the best possible for society. In other words, trying to save one resource may cause the stocks of other resources to be reduced in a way that makes society worse off. In an attempt to 'save the planet' in one dimension, the choices made by consumers driven by 'single index' information may end up making the overall resource scarcity situation worse.

Multiple resource indices?

So do the multiple resource indices do a better job? Do certification schemes such as food miles and fair trade accreditation provide consumers with meaningful signals of relative resource scarcity? Do multiple resource indices such as the ecological footprint and life-cycle assessment scores allow consumers to make better decisions?

They certainly do the job differently from the single resource-based indices, but the widening of their purview, while addressing some of the limitations of the single resource indices, opens up other complexities.

For instance, within ratings scales such as food miles there may be confounding effects. While the food miles index is designed to give consumers information on the energy demands of food production and thence the associated climate change impacts, the transportation of food is only one element of the energy story of food. For instance, glasshouses can be used to grow tomatoes through the middle of a Scandinavian winter. That would avoid their transportation from tropical climates to the south and so lower their food miles for Oslo residents. However, the energy requirements of heating and lighting the crop would more than match even the energy costs of air freight. Similarly, although perhaps not so obviously, the energy demands of growing lamb for Londoners in far-distant New Zealand are less than those incurred by local British farmers, even after including the transportation factor.[23]

Accreditation schemes face similar issues of confounding and contradiction. Ostensibly, consumers may find the information provided by any accreditation scheme useful for the decision-making process. Knowing that a brand of coffee is grown 'sustainably' and involves 'fair' compensation to growers or that fruit is 'organic' is useful to consumers for whom those product attributes are important. Producers seeing a potential to satisfy consumers with these interests respond by supplying the goods demanded. Having an independent and trusted arbiter to assure consumers that what they are buying does conform to what producers say they are supplying lowers the 'search' costs of individual buyers and so potentially improves the well-being of all involved.

The complexities to this basic story come in a number of forms. First, as with the food miles schemes, there may be internal confounding factors. Fair trade provides a useful example. What is 'fair' to one group in society may not be 'fair' to others. Paying a premium for the outputs of one group of farmers in a developing country may not be seen as fair by other competing farmers who are somehow excluded from the

---

23 Saunders, C., A. Barber and G. Taylor (2006), 'Food Miles – Comparative Energy/Emissions Performance of New Zealand's Agriculture Industry': http://www.lincoln.ac.nz/Documents/2328_RR285_s13389.pdf

'fair trade' scheme. 'Fairness', like beauty, is in the eye of the beholder.

Further confounding impacts arise when there are conflicts between the elements that comprise a certification scheme. For instance, sustainability in its strictest form requires that all three aspects of sustainability are protected: Financial, social and environmental. However, advancing one of these aspects frequently will require detriment to another. For instance, improving the financial well-being of a group of developing country farmers may be possible through gaining 'fair trade' status. That may also be achieved through the application of sound environmental management techniques. However, it may mean that the pattern of social interactions within the group, and between the benefited group and its neighbours are changed for the worse. The question then is raised: Are the financial and environmental improvements worth the social dislocation?

A trade-off across the three aspects of sustainability has to be made. That poses a difficulty for those seeking to determine if the change is sustainable and hence worthy of 'fair trade' or other sustainability-based accreditation: How much financial improvement is needed to overcome some social disruption? Critically, the judgements of the accrediting organisation cannot be the same as those of all the people seeking to be informed by the accreditation status of a product. There will be a wide range of viewpoints on such trade-off judgements within any community.[24]

---

24 The economic dimensions of these trade-offs are often poorly understood by index creators who frequently come from science or engineering backgrounds. On one level there are the 'opportunity cost' aspects: When a choice is made, one set of opportunities are enjoyed but another set is given up. More complex aspects involve the adjustments that people make to evolving circumstances: As resources are used and become increasingly scarce through time, prices increase and consumers adapt by seeking substitutes and producers look to expand supply. Omissions of this nature are evident in Foran, B. and F. Poldy, (2002), *Future dilemmas. Options to 2050 for Australia's population, technology, resources and environment*, Report to the Department of Immigration and Multicultural and Indigenous Affairs, CSIRO Working Paper 02/02, Canberra.

The need for such judgements to be made in any accreditation scheme involving multiple dimensions highlights a further aspect of complexity. Wherever judgements are involved, there is a risk that individuals will seek to manipulate the process to secure their personal goals. Gaining accredited status, if it brings greater market share or a price premium to producers, will become a valuable thing in itself. Those who are able to supply the accreditation will therefore be prone to 'persuasion' by those wanting it. The potential is therefore that accreditation becomes a less-than-useful piece of information to consumers. Producers themselves may also seek to provide information on the environmental credentials of their products that is less than accurate.[25]

In contrast, the ecological footprint seeks to provide a very broad understanding of resource use impacts. Its intention is to capture the impacts of our consumption choices on an array of resources generally. However, the 'index' used to represent these impacts is collapsed down to a single metric – area. Profiling an ecological footprint therefore ends up reducing impacts down to a single resource dimension. However, not all resource use is space-based. For instance, agricultural production can be land-intensive or -extensive depending on the amounts of non-land inputs such as labour and technology. Reducing the impacts of agriculture down to a space index runs the risk of misrepresenting resource use overall.

This is especially critical when comparisons of ecological footprints are made across countries, each of which will have different resource endowments and hence different capacities to produce different goods. For instance, the area of land required to produce a kilogram of beef in

---

25 The unwarranted touting of environmental merit by sellers is sometimes referred to as 'greenwash', but also as 'little green lies' (http://www.littlegreenlies.com/). Because buyers have less information than sellers (a condition known as 'asymmetric information'), sellers have the opportunity to provide misleading information about their products. See: Akerlof, George A. (1970), 'The Market for "Lemons": Quality Uncertainty and the Market Mechanism', *Quarterly Journal of Economics*, **84** (3): 488–500.

Japan is vastly different to that required in Argentina. And the capacity to use land for different purposes is very different between two countries like Argentina and Japan.

This is one of the fundamental principles that underpins the success of international trade – countries with different 'comparative advantages' exchange goods and services with each other to mutual advantage. To have a comparative advantage simply means that the extra costs (in terms of what has to be given up) of producing more of a good or service is lower than elsewhere.[26] The differences in costs mean that countries specialise in what they can do at relatively lower cost. It is a principle that is missed by country-to-country ecological footprint comparisons.

It is not surprising, therefore, that heavily populated countries such as Japan would be in ecological footprint deficit. The deficit reflects the comparative advantage of Japan which is not in producing goods and services that use a lot of space in order to lower their cost structures. To force Japan into an ecological footprint surplus condition would be very costly, as tasks in which Japan has a comparative advantage (such as complex industrial manufacturing) would have to be given up in order to engage in intensive food production. This is the fallacy of trying to represent resource scarcity using a single index of space.

Life Cycle Analysis and other Multi Criteria Analysis-based approaches attempt to rectify this flaw by creating indices which are without units. This in itself releases another, albeit different, set of issues: Just how to interpret the different indices becomes difficult. While it may be possible to compare different consumption choice alternatives for, say, building products when they have been assessed using the

---

[26] What this means is that comparative advantage depends on the relative performance of the trading partners. For instance, even though you are better than me at growing both meat and potatoes, it will still pay you to specialise in producing the one you are relatively better at producing (say meat), leaving me to sell you the one I'm relatively better at producing (potatoes).

same MCA process, results from a different MCA – say, of options for transporting building materials to a site – will be non-comparable. So any trade-offs between transportation impacts and product impacts become fraught.

One of the reasons for this non-comparability is that it is likely that the two indices will be made up of different resource impacts. Different products and production processes will have different impacts on different resources. For instance, building materials made from timber may have impacts on tropical forest biodiversity whereas substitute plastics may have impacts on air pollution. Hence the indices created will be made up of different components. Just what constitutes the list of resources impacted is also not defined by any conceptual framework under MCA. The choice of 'criteria' is left to the analyst's discretion. This in itself is problematic because it opens the way for the subjective determination of 'what matters' by the analyst or by the organisations commissioning the analysis. In other words, the analyst or commissioning organisation can manipulate the index so that outcomes they want to be chosen are the ones that the index supports.

However, perhaps the most critical flaw in the logic of multiple resource impacts indices is the requirement that the relative importance of resources impacted is created by applying weightings to each individual index, and then aggregating these weighted indices into a single value. The determination of weights is another opportunity for the subjective views of analysts or commissioning organisations to influence the outcome of the process. The determination of weights is therefore particularly problematic. They may be selected by the analyst directly. Alternatively, the analyst may seek input from those interested in the results of the study. Or the general public may be consulted. While the latter would appear to be preferable from an objective analysis perspective, even it is prone to difficulties and potential bias. Those members of the public consulted may be those with vested interests or unrepresentative in other respects. If the list of criteria is not complete, then there is a further risk that the weights used will be inappropriate. For instance, if the price of product is not included as a criterion – as is the case in most LCA studies – the weights given to criteria that are very difficult to achieve (and so only found in more expensive products) will be given inordinate relative importance.[27]

Again, the selection of weights can be made so that the process of index calculation is biased toward recommending outcomes that are advantageous to the interests of specific interest groups.

How to account for resource scarcity in consumption choices?
The conclusion to be drawn from the analysis of single resource indices is that they are so limited in their scope that if consumption decisions were based on one of them alone, the overall resource scarcity picture may be made worse and that if more than one were to be used, contradictions could leave the consumer in a state of confusion.

The conclusions to be drawn from the consideration of multiple resource indices as a basis for consumption choices is that they are potentially internally contradictory, sometimes conceptually flawed and offer the potential for vested interests to take advantage of them to secure outcomes they want.

Both single and multiple resource indices are also costly in themselves to implement. A new industry has developed to perform life-cycle assessments, carbon accounts and ecological footprints while certification authorities and organisations have also proliferated. It is important to recognise that these activities are also responsible for using scarce resources. People who are employed as 'carbon accountants'

---

27 An example of this problem comes from the *Economist* newspaper's ranking of the most liveable cities of the world. Cities such as Melbourne, Vancouver and Vienna vie for top ranking, but the cost of living is not included as one of the ranking criteria. With seven of the top ten ranked cities in Canada and Australia, it is clear that the weightings applied to criteria associated with low population densities were especially heavy. The process used to select the weights is not disclosed (see: http://www.economist.com/node/21528162).

to calculate the life-cycle emissions associated with the full array of products on the market, for instance, could be employed in other ways to the benefit of society. The question, then, is whether society gains more from their outputs than it has to pay in costs. An important part of that answer is knowing what else is available to perform their role.

If both forms of indices are so problematic, what can consumers rely on as a means of learning about the impacts of products on relative resource scarcity? The answer lies in the signalling role that the market price of goods and services plays. When suppliers bring a product to market, they have had to assemble a range of resources to create the saleable item. Every time a resource has been used, the former owner of that resource has had to be paid at least as much as they hoped to achieve from their use of the resource. In that way, the cost of each resource represents the benefits forgone when the next best use of the resource is given up. When a resource is relatively scarce, it means there are a lot of other uses competing for it. That means its cost will be higher. When a resource is relatively plentiful, there are few other competitors seeking access and its price will be lower. In this way, the prices paid for resources used, and hence the costs of the producer, will reflect the relative scarcity of all the resources used. The price paid for the product, by reflecting the costs of production (the price paid has to be at least as much as the costs of production to make it worthwhile for the supplier to bring it to market) therefore reflects the relative scarcity of the resources combined in the product.

In this way, the price of a product is the measure of the scarcity of the resources used in its production. In a sense, therefore, it is a single 'index' of resource scarcity. Only if a potential buyer values the product more highly than the combined 'next best use' values of the resources used will a purchase be made.

If price acts so well to signal relative resource scarcity, why have so many people found it to be so unsatisfactory as to require the wide array of alternative 'indices' that have been set out in this chapter?

One reason is that resource scarcity is not the only thing that indices signal. Many of the accreditation schemes that have been identified in this chapter as having a resource scarcity signalling role are also attempting to demonstrate other aspects of consumption choices. For instance, organic status food is seen by some consumers as a health signal. Others see fair trade as an indicator of the impact of their choices on the well-being of poor people in developing countries.

Another reason for the proliferation of indices is that some of the dimensions of resource scarcity that are targeted by the indices are seen by their users to be outside the purview of the market and so not reflected in market prices. For instance, some of the dimensions of the environment included do not have well-defined or well-defended property rights and hence are not bought and sold in markets. Without exchange, prices cannot form to reflect relative scarcity.

The first response to this 'defence' of indices must be to make sure that property rights to environmental assets are well-defined and properly defended so that prices can form to reflect relative scarcity.

There are cases, however, where property rights to environmental assets are so costly to define and defend that it's best not to move in that direction. Yet even then policy measures can be designed on a resource-by-resource basis to make sure that price signals come to the fore. For instance, the levying of taxes on environmentally damaging emissions add to the costs of production and are thus reflective of the environmental costs caused by consumption activities. In this way, product prices signal environmental as well as other resource use. But this logic should not be taken as a green light for any and all government intervention. Only interventions that can be demonstrated as beneficial to society should be implemented. That can best be judged by a consideration of all the costs and benefits a policy causes, including the costs of setting up and administering the policies.

Prices as they stand now do not perform the scarcity signalling role perfectly. The failure of many governments to instigate and then enforce policies that address environmental resource use issues appropriately is to blame. Indeed the difficulties of designing and implementing such a suite of policies means that prices will never be able to perform their resource scarcity signalling role perfectly. Perfection comes only at a prohibitively high cost.

However, prices, imperfect as they are, have the advantage of embodying the knowledge inputs of everyone who engages in buying and selling. They are formed independently from any vested interest groups. These two key features set prices apart from indices as superior signalling devices.

In some circumstances we know for sure that prices are far from perfect. It may be that trade restrictions imposed by governments to protect local industries or a lack of competition between suppliers causes prices to give the wrong signals about relative values. In other cases, some governments may impose policies to manage the environment while other governments don't. Prices in one country will reflect environmental scarcity but won't in another. Consumers are then more likely to choose the cheaper products from the country without environmental protection policies.

Do these price imperfections mean that we should give up on prices and instead rely on indices and more government policies to restrict trade? Not at all. Such measures would serve only to make the resource allocation process worse because they open up the prospect of political manipulation. Vested interest groups would use the chance to have their ways. Rather, the preferred policy approach is to work toward removing as many of the price-distorting factors as possible. For instance, barriers to competition both domestically and internationally should be dismantled, the citizens of rich countries can pay the costs of improving environmental management practices in poor countries and government subsidies paid to environmentally harmful activities should be scrapped.

It's also worth noting that the capacity of prices to act successfully as signalling devices is likely to improve over time with technological advances.

One problem with the price signals that are currently available is that for some consumption decisions, they are not very timely. For instance, electricity prices are generally not so flexible as to indicate the changing relative scarcity of resources across the course of a day. Peak demands cause more scarce resources to be used and so the price should rise to reflect this. Most supply systems don't send this type of signal, mostly because the costs of metering use and relaying changing prices have been too high to warrant it. But with 'smart meters' being installed in many places, people are being able to watch their energy use and the price they are paying per unit as it happens. Advancing technology will therefore allow prices to become better at performing their signalling function.

Similarly, there are many advances possible in water supply pricing. In dry times, as reservoirs are run down, the scarcity of the water resource increases. Yet most utilities charge a standard per kilolitre price that is invariant over seasons. Increasing prices (and announcing those increases widely) as storage levels fell would provide much more accurate signalling of relative scarcity. Price regime reforms such as the introduction of seasonal pricing are possible immediately, yet political pressures to hold down the price of water (as a necessity to life) prevail. The irony of this is that stopping prices from signalling scarcity simply means another form of rationing needs to be installed. This is usually a quantity regulation approach that can cause many inefficiencies, including the 'rationing by ordeal' such as is involved when older people are forced to water their gardens by carrying heavy buckets full of water because of restrictions upon watering by hose.

The approach of relying on prices to signal resource scarcity is the thrust of the 'getting the prices right' argument that has been developed in the 'sustainable consumption' literature.[28] Much of that literature has involved calls for government-imposed restrictions to be placed on the volume and type of consumption people choose, with justification drawn from indices such as the ecological footprint. The restrictions upon individual freedoms such impositions would involve to one side, they would also reduce the well-being of people as they would not be permitted to pursue their preferred consumption options. The determination of which goods and services to restrict and by how much would face the same types of complexities posed by the single and multiple resource scarcity indices raised earlier in this chapter. The prospect of subjectivity and hence corruption would also be real. Instead, the argument is advanced that if prices are allowed to perform their signalling role and they are determined in circumstances where appropriate environmental protection policies have been established, then the sustainability issue of making sure resource scarcity is accounted for in people's consumption choices will have been addressed.[29]

A 'little green lie'?

Concerns held by well-intentioned members of society for the 'sustainability' of our civilisation have to some degree been responsible for the proliferation of non-price resource scarcity indices. In fear that the consumption decisions made by others are driving the whole of society on a trajectory that they see as being dangerous, they have sought to influence consumption trends with measures that go beyond what they see as the failed process of market rationing by price. What this chapter has sought to demonstrate is that the non-price alternatives are themselves problematic – to the point where they may provide signals that are not in the best interests of society at large. In many cases

---

28 Bennett, J. and D. Collins (2009), 'The policy implications of sustainable consumption', *Australasian Journal of Environmental Management*, **16** (March): 47–55.
29 This can also be extended to include social issues, with the distribution of income being accounted for through the taxation and welfare distribution systems.

this is because the task that the indices try to achieve is so complex that they confuse the picture of scarcity that they are trying to deliver.

The danger here is that the well-intentioned, seeking to help others in society, end up creating a worse situation than otherwise would be the case. The 'little green lie' for these people is seen as a representation of the worrying truth.

Even more dangerous is the prospect of 'help' being given to both producers and consumers. For instance, suppliers of tropical timbers may be required to submit to index-based production constraints and then consumers may also be 'assisted' in their consumption decisions by a separate array of indices relating to purchases of timber-derived products. Regulations may be imposed on energy suppliers to restrict their greenhouse gas emissions and then consumers given a set of directives regarding their energy use. These would represent instances of 'double counting' environmental impacts.

However, there are also those who can gain reward from the promulgation of the 'little green lie'. Those who would be advantaged by the changes in choices made by consumers adopting the logic of the indices could be expected to be supportive. First amongst these are the producers of goods that the indices are showing to be favourable to the future of the world. Local farmers who have 'low food miles', renewable energy generators who lower the carbon footprint, suppliers of bicycles and public transport, etc. are likely advocates.

So, too, are those involved with the 'indices industry'. The consultants who perform life-cycle assessments, the carbon accountants, those who provide the certification for sustainable tropical timbers, etc. all enjoy a source of income thanks to the demand for indices in their various shapes and sizes.

This picture is a familiar one that has been cast as the 'Baptist and the bootlegger' parable.[30] In the prohibition era in the United States, the

---

30 Yandle, B. (1983), 'Bootleggers and Baptists: The Education of a Regulatory Economist', *Regulation* 7 (3): 12.

members of the religious right (represented by members of the Baptist Church) were chief in supporting the banning of the sale of alcohol. These were well-intentioned members of society seeking to make others confirm to a behavioural standard (alcoholic abstinence) that they believed to be in the interests of society as a whole. Whole-hearted support for this policy was provided by the bootleggers – those people who became very wealthy through their activities in producing and distributing illicit alcohol at black market prices. Strange 'bedfellows' emerge in public policy deliberations.

4: Population

Proposition: World population should be capped.

More people mean more pressure on the world's scarce resources, including the environment. The only way to protect the environment, stop starvation and to ensure that there are enough resources for future generations is to stop population growth.

BUT

People are a resource. They have the capability to develop innovative technologies and institutions to deal with growing scarcity in specific resources. New ways to satisfy people's wants and new sources of scarce resources can be discovered. If limits are installed, who will they affect?

Too many people?

Resource scarcity was the focus of the previous chapters' 'little green lies'. The focus has mostly been on the amount of various scarce resources remaining available into the future to satisfy people's needs. One of the key factors influencing resource availability (and scarcity) is of course the rate at which resources are being used. That in turn is driven by numerous forces, principal amongst which is the size of the population drawing on the stock of scarce resources. It is not surprising, therefore, that associated with concerns about the scarcity of resources would be worries that the population of the world is already too large and, furthermore, is growing too quickly.

Hence, along with estimates of the 'days left' of oil, coal and other scarce resources, the website 'worldometers'[1] also reports the number of people inhabiting the planet (passing the seven billion mark on 31 October 2011). The number of births for the year is also tracked and shown to be increasing at a much faster rate than the number of deaths for the year.[2]

'Worldometers' estimates the population of the world by applying the current growth rate of 1.15 per cent. The mathematics of this operation results in increases in population that 'compound' over time. This is the process that worried the Rev. Thomas Malthus[3] as far back as 1798. Malthus saw an apparent conflict between the rate at which population grows (a compounding process) and the rate at which he could see food production could grow (a simple linear, non-compounding progression). He concluded that, eventually, the demand for food would outstrip the capacity of the earth to ensure supply. Starvation was the predicted result. Similar predictions were made by Paul Ehrlich[4] after he visited the crowded cities of India in the 1970s and the tradition of 'doom-saying' continues to fuel the population debate today.[5]

There are, however, some key points of finer detail to note about the current growth rate of world population. The first is that it is now almost half of the highest rate of population growth (recorded in 1963 when it reached 2.19 per cent) and the trend remains downward. 'Worldometers' is predicting that it will fall to less than 1 per cent by

---

1 http://www.worldometers.info/
2 These trends are likely to be alarming to those who enjoy the tranquillity of a deserted beach but reassuring for those who like the hustle and bustle of a crowded city street. Someone born and raised in Shanghai is likely to feel uncomfortable if left alone in an Alaskan wilderness. Similarly, the experience of a day in the hordes of Calcutta may be intimidating to a person accustomed mostly to their own company living in outback Australia. As with most things, preferences are relative.
3 Malthus T.R. (1798), *An essay on the principle of population*. Oxford World's Classics reprint.
4 Ehrlich, Paul R. (1968), *The Population Bomb*. Ballantine Books
5 For example, see: Brown, Lester (2009), *Plan B 4.0: Mobilizing to Save Civilization*. New York. The Earth Policy Institute.

2020 and to less than half of one per cent by 2050. The UN[6] forecasts that this continual decline in the growth rate will cause a plateau in the world population at around nine billion some time between 2050 and 2075.

The second point to note is the great variety of population growth rates evident across countries. At the high end (over 3 per cent) of the growth rate charts[7] are countries such as Burkina Faso, Ethiopia and Zambia. At the other extreme, countries with negative growth rates include Japan, Estonia and the Czech Republic. What this indicates is that the bulk of world population growth is being generated in developing countries. Most developed countries have low or negative natural population growth rates (the growth rate net of immigration impacts).

The trend toward lower population growth rates and hence the eventual stabilisation and even decline of the world's population is linked to the level of development achieved and the rate at which it is achieved in the developing countries of the world. The irony for the resource scarcity picture, therefore, is that population will eventually stabilise but only because incomes and hence consumption rates increase in developing countries. This has given rise to a debate around just what population the world is able to support: What is the world's 'sustainable population' level?

Some, such as EvFit,[8] argue that the figure is as low as 600 million. This is on the basis that the use of fossil fuels has allowed the population to grow to what it is today. They argue that once non-renewable fuels are no longer available, we will need to go back to the levels of population evident in the early 19th Century. Others argue to the contrary that any downturn in population growth rates will result in aging populations

---

[6] UN (2004), *World population to 2300*: http://www.un.org/esa/population/publications/longrange2/WorldPop2300final.pdf
[7] https://www.cia.gov/library/publications/the-world-factbook/fields/2002.html
[8] http://www.evfit.com/population_max.htm

that will have difficulty in supporting themselves.[9]

Those who are concerned about the overall size of the population and its continued growth are often advocates of specific policy measures that are designed to limit population size and growth. For individual countries, particularly in the developed world, these often centre on reductions to the rate of immigration.[10] For the whole world, however, such strategies are futile. For the broader context, mandatory birth control and family size limitations are advocated. Examples of such policies are found in China where, since 1979, parents of Han ethnicity are permitted to have only one child on risk of heavy fines or forced abortion and where compulsory sterilisations have been reported.[11] India also had a policy of forced sterilisation between 1975 and 1977 which, due to its widespread unpopularity, was replaced by a national family planning initiative that continues today. The China one-child policy is similarly facing pressure for reform. This is being generated by concerns that single children will have increasing difficulty in meeting the welfare needs of aging parents.

While there is debate about estimates of future population levels, there is even more heated dispute regarding the appropriate future 'management' of the world's population. Elements of that debate extend beyond the environment and into religion. But what does need to be kept in mind is that the debate is about people and their choices to have children. That is a very personal choice and one that is central to most people's lives.

Population controls

One of the lessons learned from the implementation of mandatory population control measures is that they can be inordinately unpopu-

---

9 http://longevity-science.org/Population_Aging.htm
10 http://www.population.org.au/
11 http://www.independent.co.uk/news/world/asia/chinese-state-holds-parents-hostage-in-sterilisation-drive-1947236.html.

lar and have severe unintended consequences. Their impacts are profound not just immediately on those people who are forced into family decisions that they would otherwise not choose. They can also affect whole societies. For instance, the China one-child policy is directly responsible for higher rates of female infanticide and subsequently for an imbalance between the sexes in the current child-bearing generation. Dissatisfaction amongst men who are unable to find partners is apparent in criminal activities such as smuggling young girls as brides-to-be from neighbouring countries.

To achieve targets such as the 600 million espoused by EvFit, it is clear that draconian population growth rate controls would be required. It is even implied that population reductions would be welcome. No doubt policies to reduce the population would be even more unpopular than those required to reduce the growth rate. Both types of policy require an answer to the same question: Who would be the subject of such policies? Clearly those espousing the idea of reducing population size are not volunteering to help the cause by committing suicide,[12] but in the end those seeking a smaller world population will require significantly changed behaviour from people of reproductive age. A forced departure from their otherwise intended behaviour will be unpopular.

This unpopularity indicates that the imposition of such policies certainly reduces the well-being of those who are directly affected. The question must be whether the broader social impacts of population controls are worth these costs.

The argument for the imposition of the one-child policy in China was that it was necessary to avert the reoccurrence of the 1962 famine that caused some 30 million deaths.[13] However, the relationship between population levels and famines needs to be considered more carefully

---

12 While not advocating suicide, the Voluntary Human Extinction Movement advocates the cessation of procreation to cause the eventual extinction of the human species: http://www.vhemt.org/
13 http://www.time.com/time/world/article/0,8599,1912861,00.html

before reaching that conclusion. The real cause of China's inability to feed its people in the 1960s was the implementation of communal ownership and food self-sufficiency requirements. The reforms initiated by Chairman Mao Zedong, founding father of the People's Republic of China, involved communal ownership of the land resource and the formation of large communal farms. Without the incentives provided by reward for individual achievement, the productivity of these farms was very low. Furthermore, people were relocated from more fertile coastal regions to the western regions in an attempt to spread the population more evenly across the nation. The imposition of food self-sufficiency requirements at a provincial level, meant that the less agriculturally productive inland areas had to feed more people from local sources. The outcome was starvation. It was not brought about by the weight of population numbers but rather by the poor resource management structures put in place by the Chinese Government.

Relaxation of trade restrictions between provinces made some impacts on food availability and hence absolute starvation rates, but it was not until 1979, when the Household Responsibility System (HRS) was introduced, that the pattern really started to change. Under that system, farmers were given 'use rights' to land they occupied. So while they did not own the land itself, they owned any produce grown on the land. This single reform allowed the productivity of land to rise so that farmers not only grew enough to feed themselves but also produced a surplus that they could sell, for personal gain, to people living in the cities. International trade in agricultural products has further permitted over 400 million Chinese people to be lifted out of absolute poverty and to avoid the catastrophic circumstances of the 1960s despite the almost doubling of the nation's population in the interim.

The implication of the Chinese story is that it was not population controls that saved the nation from starvation. Rather it was the abolition of policies that impaired the incentive for, and capacity of, people to use available resources to best advantage and their replacement with policies that proved successful. In this way, people became the solution to the problem rather than the source. First, people were able to develop and implement social structures that worked to deliver incentives for production and exchange. Second, people were able to deliver food production sufficient to satisfy the demands from across their own nation and from abroad.

People are the answer
The somewhat paradoxical argument that people are the solution to the population issue underpinned the stand taken by Julian Simon[14] in his opposition to the doom-saying predictions made by Paul Ehrlich and his followers.

A clear way of understanding the paradox is to contemplate why a world population of 600 million is far less than what can be supported by the earth. Those advancing the 600 million figure argue that it is the level of population that can be supported without the use of fossil fuels: Their argument is that because the world will run out of fossil fuels, the population should not be any greater than this level. What the argument ignores is the reality that humanity is not solely dependent on the existence or otherwise of fossil fuels. The reason for that is the development of other sources of energy. Technically, it is already possible for humanity to maintain its current energy consumption patterns at its current population size without the use of any fossil fuels. Of course it would require a dramatic transition from fossil fuel to, say, nuclear power and away from petroleum-based transportation to electric vehicles, but technically it could be done. Over time, technological advances will continue to make such a move out of fossil fuels easier and cheaper. How has this occurred and how will it continue to occur? The answer is through the inputs of people. The ingenuity of people – from their ideas through to innovation in the production of manufactured

---

14 http://www.juliansimon.com/

capital (for example, nuclear power plants) and the development of social capital (for example, private property rights) – has meant that the productivity and availability of natural resources has improved significantly over the past 200 years and will continue to do so.

A useful analogy is found in agricultural productivity. In the United States[15] over the period 1946 to 1998, the productivity of a unit of land increased on average by nearly two-and-a-half times. In other words, over that 52-year time period, the same parcel of land could support the food needs of more than double the number of people. Those productivity gains did not come from additional labour inputs (the number of people employed fell to around 30 per cent in 1998 of what they were in 1946). They came from improved farm management knowledge and new breeds of plants and animals (the products of human capital), mechanisation, pesticides and herbicides (the results of human capital being applied to the production of manufactured capital). To take another example, in Vietnam, the output of rice from land has increased at an even greater rate through the application of human and manufactured capital but also the reform of social capital. The property rights reforms undertaken by the Vietnamese Government whereby private use rights have been given to farmers over the lands they have worked (even though the land remains under state ownership) have been primarily responsible for the improvements observed.[16]

The Vietnam rice story along with the results of the introduction of the HRS in China also demonstrate the importance of incentives in ensuring that natural resources are used to their best advantage and that, over time, humanity adapts to changing patterns of resource scarcity. In China and Vietnam, it was only when individuals were given the responsibility (and opportunity) to make themselves as well off as they could be from the resources they had available, did agricultural

---

15 See United States Department of Agriculture, Economic Research Service: http://www.ers.usda.gov/publications/arei/ah722/arei5_1/arei5-1productivity.pdf

16 See Pingali, P. and V. Xuan (1992): http://www.jstor.org/stable/1154630

productivity lift. When the collective system of communal ownership and equality of outcome regardless of effort was in place, the incentives for individual betterment were diminished. Without that incentive, there was little reward for effort and the outcome was poverty. The creation of private property rights (at least to the products of their labour) and the provision of a social setting in which trade between people could occur enabled productivity improvements and adaptation to both the changing patterns of demand for their products and the availability of the inputs necessary for the production process.

Once trade was possible, prices could form and were able to start performing their role as relative resource scarcity signals. With the opportunity for betterment as their goal, farmers looked to find ways of reducing their costs relative to the output they could generate from using resources.[17] They also sought to produce goods and services that people wanted rather than simply satisfying state-imposed quotas on production, which more often than not were either less than what was demanded or left to rot unwanted. The way they knew what to produce was through the signals sent through the marketplace as prices.

The constant adaptation of people to changes in prices was the foundation of Julian Simon's success in his famous bet with Paul Ehrlich. The two bet on the price of a bundle of what Ehrlich defined as key indicators of the world's scarce natural resources. Ehrlich predicted that their collective prices would rise over a pre-assigned time period as they became more and more scarce. Simon said the prices would be stable or fall because, if their prices did rise, people would adapt to those price rises by substituting away from them, discovering more of them, or developing alternatives for them. The Simon contention was that human ingenuity would outpace natural resource scarcity. Simon

---

17 Note that this is not the same as minimising costs. To maximise wellbeing, a farmer may choose a production operation with higher costs if its outputs are greater or worth more. The object is maximisation of net profits, not minimisation of costs or maximisation of revenues.

won the bet: The prices of all the resources in Ehrlich's basket fell in real terms over the agreed period.

Importantly, the institutional structure or social capital needed to see markets work and for people to have the incentive to make the necessary adaptations and developments had to be in place for Simon's position to be fulfilled. This is an important aspect of population growth in itself.

More people?

Under the Simon logic, it would appear that more people is a good thing for the world: More people means more prospects of ideas and innovation that will allow adaptation to natural resource scarcity. An important caveat to this conclusion is that currently a large part of the world's population play little role in the adaptation process and have little prospect of doing so. This is mostly because the social institutions of the communities in which they live do not provide strong incentives to act, or actively limit their capacity to act, because their levels of human capital (education) are so limited. So, while peasant farmers in China adapted to increase their productivity when the HRS was introduced, the rates of growth achieved have not been as great as they could have been (and could still be) under a social institutional structure of private land ownership. Similarly, the productivity response by rice farmers in the Mekong River Delta region in Vietnam could have been even greater if the education levels of farmers were higher.

Similarly, the potential of millions of highly intelligent people in developing countries to become innovators, inventors, in short, resource scarcity problem-solvers, is simply lost because of their lack of a formal education and other barriers to the development of their talents.

The question, then, is not just one of the number of people affecting the likelihood of adaptation to natural resource scarcity. It is also a matter of the capacity of people to fulfil their potential. That is in turn dependent on the social institutions that are in place.

The adaptation process involves people making consumption decisions which recognise natural resource scarcity through the signals that prices send. People also make choices about their well-being and how it is affected by the number of children they have. It is useful, therefore, to consider the reproductive choices made by people when understanding overall population levels. What level of population is going to make people best off?

Population choices
Having children is a matter of choice for a large proportion of the world's population of reproductive age. Birth control is now mostly well-understood and available at relatively low cost. However, that is not true for all. Social barriers to contraceptive use remain in many societies. Up to 40 per cent of pregnancies in developing countries and 47 per cent in developed ones are argued to be 'unintended'.[18]

In the first instance, it is useful to consider pregnancies that are a matter of considered intenion. To analyse the choice of having a child it is instructive to take a step away from the emotions that drive the decision. This is a matter of going beyond the choices made by individuals and instead looking at decision-making patterns we can see throughout society. These patterns will not explain every choice of every parent to have a child, but their analysis does help to see through the complexities of individually specific choices. They help to provide a better understanding of the key variables that drive people and hence what affects their well-being.

Put in its most basic form, the choice to have a child involves – amongst a host of emotional drivers – people weighing up factors that can best be described as costs and benefits. The reproductive logic then becomes: Only if the benefits of an extra child outweigh the costs will people go ahead and choose to reproduce.

---

18 Engleman, R. (2011): http://www.thesolutionsjournal.com/node/919

On the positive side, the benefits of having a child are diverse and numerous. But three are of interest. First, a child is a potential source of labour to the family enterprise. So, for a farming family, an extra pair of hands available throughout the annual cycle of planting and harvesting the crop, another person to tend the family's animals, to gather firewood or walk to the well to collect water is a valuable addition to the resources available. This logic doesn't just apply to farming: It is equally evident in any family-based small business, be it in the light manufacturing or service sectors. Labour is valuable.

Second, children are also valuable to parents when they need to be cared for in their old age. A child can earn the income necessary to meet the needs not just of themselves but also their aged parents. That care goes beyond the provision of goods and services that can be purchased.

The third benefit of children relates to risk. Throughout a child's life there are risks faced by parents that the benefits of having the child will not be realised. The child may die at birth or in its early years. There is also the chance that a child will not remain with the family or indeed offer any support to the family as it reaches adulthood. A key way of dealing with these risks is for parents to 'diversify their portfolio' by having a larger number of children. The probability that they will secure an able and willing supply of labour for their enterprise and for their old age care is increased as the number of children increases.

These and other benefits of having children are weighed up against the costs. These too are wide-ranging and vary from case to case. Again, three cost-types are of general relevance here.

First, with increasing family size comes an increase in the cost of housing. Clearly, housing demands are different in different cases and the requirement for more space with more children can be 'lumpy'. This means that space requirements may not increase smoothly with each additional child. An existing house may be satisfactory for three children but a fourth child would mean the purchase of a bigger house or the expansion of the existing house.

Second, a component of the costs of an additional child relate to education. These costs are not only the payment of school fees, books, and equipment but they also include opportunity costs. If a child is spending time attending school, then they are not able to engage in their next best use of that time, which may be working in the family enterprise. Hence a cost of school time is the forgone value of that time spent in the family business.

The final general cost factor involves the care of the child. A pregnant woman approaching the time of birth may be restricted in what she is able to do. She may have to stop working in the family business or as a salaried employee. That change may also continue for some time after the birth, depending on her recovery and the condition of the baby. Once the mother and baby are sufficiently recovered from the birth, parenting begins and may involve mother and/or father. No matter what the mix of parenting roles, it remains the case that changes to the established pattern of activities will occur. Either or both the mother and the father will be taken away from other activities to care for the infant. All of these activity changes mean costs. These are the opportunity costs of the activities that are given up in order to deliver the baby and then care for the child through infancy.

Balancing these benefits and costs of having children will necessarily produce different outcomes for different people. However, consideration of the three benefit elements and the three cost factors leads to some useful conclusions regarding trends in the decision to have a child as economic prosperity increases.

First, consider what happens to the demand for children as people living in more developed economies become better off. The benefits of having children as a labour source declines because children are required by law to attend school and other laws prevent children under specified ages from working. Parents thus face the prospect of legal action if they seek to use their children's labour.

With respect to children acting as aged-care sources, as development progresses, a number of substitute sources of care emerge. With more efficient credit markets (primarily due to well defined and defended property rights over bank deposits) people are able to save through their lifetimes in order to pay for the delivery of services in old age. Pension plans and superannuation schemes, both publicly and privately supplied, support the facility of saving for future consumption. With greater security provided through savings schemes, children are less 'competitive' as suppliers of aged care.

Finally, a key element of the risks involved in having children in order to supply future services (both labour and aged care) is infant mortality. Rates of infant mortality are strongly correlated with economic growth. With greater wealth comes greater capacity to invest in health care (both privately and through the public sector), notably both of mothers during pregnancy and babies in the first months of life when they are most vulnerable. Evidence of this relationship is found in data on the number of deaths per thousand live births.[19] From Angola (175.9), Somalia (105.56) and Nigeria (91.54) at the high end, through to France (3.29), Japan (2.67) and Singapore (2.32) the contrast is stark.

Changes in all of three 'benefit' factors indicate that, as economic development progresses, the demand from couples to have children will fall.

On the cost side of the child choice equation, development also has significant impacts. With greater development, comes more urbanisation. The costs associated with more living space for more children are greater in urban settings than in more sparsely populated agricultural areas.

The costs of educating children also increase. In part this is because with development comes greater parental aspirations for their children. This means more years at school for children as well as greater demand

---

[19] https://www.cia.gov/library/publications/the-world-factbook/rankorder/2091rank.html

for higher quality in the educational experience. That comes at a cost.

Finally, the opportunity costs of child-bearing and child-raising are potentially the most significant cost factor. With more development and more opportunities emerging, particularly for women, the costs associated with time out of the paid work force to raise a family increase.

Together these three factors combine to create a situation where increasing economic wealth causes the costs of bearing and rearing children to rise.

Bringing together the benefit and cost factors produces a reinforcing effect. With benefits falling and costs rising, it becomes less likely that people will want to have more children. The evidence of lower population growth rates in developed countries relative to those witnessed in the developing world substantiates this prediction. It is apparent that, as people become better off and live in societies that are better off, they want to have fewer children. Furthermore, as people become better off they are also better able to afford contraception and are more knowledgeable about its use.[20]

Implications for policy

People around the world strive to make themselves and their communities better off economically. It would also seem clear that with more economic development people show a desire to have smaller families. The key message, then, is that people would be both better off and the world's population would trend toward negative growth more quickly if economic development could be spread more widely across the world.

There is an irony, then, to the calls for population growth controls. It is not so much that people have an inherent desire for more children

---

20 The data suggesting that the rate of unwanted pregnancies is higher in the developed world than in developing countries indicate that the rate at which the desire to have children falls even faster than the rate of adoption of contraceptives as development progresses.

that has to be somehow restrained. It is more the case that they have larger families because of the conditions of poverty in which they find themselves. If that poverty could be alleviated, they would choose not to have as many children. It would not be necessary to impose population controls. Those controls would be voluntary and life-enhancing.

The trends that go along with development in terms of improving educational standards and better health care also support the Simon stand on the importance of human capital in the constant adaptation to changing relative resource scarcity. A population that is better educated and in better physical health is more likely to be innovative and adaptable. Ignorance and poor health are ingredients for inflexibility and an inability to move out of the cycle of poverty. Growing resource scarcity (caused in the case of farming land, for instance, by increasing soil erosion and salinity) in such cases leads to growing poverty rather than an adaptive response away from land-use practices that have caused the resource to degrade.

It is also clear that social institutional measures recommended to ensure adaptation to resource scarcity are also measures that have proven successful in stimulating economic growth. Prices were advanced as an effective way of signalling relative resource scarcity to buyers and sellers in markets. The incentive of self-improvement and the operation of prices in markets, given their foundation in well-defined and defended property rights to resources, have been shown to ensure adaptation to changes in relative resource scarcity. Those very same structures have been demonstrated across the world as being the necessary ingredients for ensuring the best prospects for economic growth.

The policy picture that is emerging is clear. Making sure that property rights are well-defined and defended and that people are free to pursue their individual happiness through the buying and selling of goods and services in competitive markets will generate a number of key outcomes. Economic growth will be enhanced and people will be adaptable to changing conditions of relative resource scarcity. And people will be able to make choices about their preferred family size: Choices that will see world population growth rates fall.

One of the more prominent ways in which societies lose well-defined and defended property rights is through war. Even if the legal structure of a community survives a war, the means of enforcing the rule of law are frequently lost. And with outright invasion, existing titles to property are usually lost completely and replaced with an alternative vestment of rights. The process of resource allocation through competitive markets in such circumstances is highly disrupted. In particular, long term planning for resource use is lost. As a result, countries engulfed in war, civil or otherwise, are predominant in the list of countries afflicted by poverty and starvation. Facing the risk of having their crops destroyed or seized, farmers will be reluctant to invest in planting. There is also the potential for war to spark an acceleration of population growth given the additional risks imposed by war that children will not survive to be supporting adults. War therefore presents a real challenge to the capacity of the human system to adapt and survive.[21]

Unintended population growth

The analysis presented on choices regarding the number of children people have is an insufficient treatment of the topic of population growth because it fails to deal with unintended and unwanted pregnancies. By definition, these are events that decrease the well-being of the parents. If the intention was not to have a child and the pregnancy is not welcomed then well-being will be diminished.

Unwanted, unintended pregnancies arise for any number of reasons. The relative expense of contraception is one. However, social institutions are also important factors. In societies where contraception is not accepted for religious or other socially derived reasons, the chances of

---

21 For instance, the fertility rate in Timor Leste increased to around 7 after the cessation of the conflict there over the nation's autonomy from Indonesia as families grew to make up for the decline in births during the war and the loss of youth in the fighting.

more children being born than are wanted are clearly higher. Societies where women are unable to assert their child-bearing preferences are also likely to experience higher birth rates. This is especially the case where women are limited in their exposure to the world outside the immediate household due to social convention. In such cases, the opportunity costs of women's work outside the home is zero. Hence the family experiences no relative loss of earning power through the woman staying at home and bearing/rearing more children.

It is therefore not only economic growth that reduces population growth rates. Social institutions are also potential areas for change that affect population growth.

There is some evidence, however, to suggest that economic growth and social institutions are linked. This comes in the form of total fertility rate data across different countries. These data report the number of children per woman across nations. Any rate less than 2.1 represents a situation where insufficient children are being born to replace their parents. As would be expected from the previously reported data on differential population growth rates, fertility decreases as the level of economic development increases. However, other trends associated with social institutional factors are not so apparent.

For instance, one social institution often criticised by advocates of population control is the Catholic Church. This is because of the Church's adherence to a policy that prohibits the use of contraception. One would therefore expect to see countries where Catholicism is the dominant religion with fertility rates well beyond the self-replacing level of two children per woman. The data show otherwise.[22] For instance, Italy has a fertility rate of 1.39. Spain's rate is 1.47 while Poland stands at 1.3. All of these rates are lower than China's one-child-policy-driven level of 1.54. At the other end of the fertility spectrum it's difficult to

---

22 https://www.cia.gov/library/publications/the-world-factbook/rankorder/2127rank.html

find a Catholic-dominated country. Typically there are the sub-Saharan nations with Niger (7.6), Uganda (6.7), Somalia (6.35) and Burundi (6.16) being among the leaders.

It seems that even the strength of social institutions regarding contraception can be challenged by the power of economic conditions to change people's choices. That is not to say that social institutions are irrelevant. For instance, predominantly Muslim counties have fertility rates that are consistently above the threshold of 2.1: Jordan (3.39), Pakistan (3.17), Libya (2.96), Syria (2.94) and Indonesia (2.25). As the most populous Muslim state, Indonesia can be compared against neighbouring (Buddhist) Thailand with a fertility rate of 1.66. The difficulty of drawing conclusions from these data is, however, that these countries have differing levels of economic wealth. Hence, it is difficult to determine if social institutions or differential economic wealth, or a combination of the two, are behind the differences in observed fertility rates.

Perhaps of greater significance is the observation that countries with the lowest fertility rates are those with well-developed economic systems and high population densities. Singapore (1.11), Hong Kong (1.07), Taiwan (1.15) and Japan (1.21) have amongst the world's lowest fertility rates. With so many people in close proximity, it would seem that people have less demand for more of them. When fertility rates fall to these levels, 'birth-dearth' sets in and population decline appears to become irreversible.[23]

A 'little green lie'?

The issue of population growth and associated calls for controls are vexed. Many people hold very strong views on the topic. Those views are often driven by deeply held religious beliefs or passionate concerns

---

23 Ron Duncan and Chris Wilson (2004), *Global Population Projections: Is the UN Getting it Wrong?* Working Papers in Economics and Econometrics No. 438, Australian National University, Canberra.

for the future of the world. The topic is difficult, too, because choices regarding children are amongst the most sensitive anyone makes. They involve the lives of people, not in the abstract but in the flesh.

The calls for population controls are unpopular because they would force outcomes onto people that they otherwise would not choose. The question is therefore whether the disquiet likely to arise from population controls is worth the perceived benefits arising.

The argument advanced in this chapter is that population controls are unnecessary. Paul Ehrlich's population bomb has not yet exploded and is unlikely to do so, so long as some key processes are in place. Principal amongst these is the operation of markets and prices to signal relative resource scarcity to a population that is capable of – and has the incentive to – adapt through innovation. It was through the actions of inventive adaptable people that Julian Simon saw the population bomb being defused. It is also the process that will ensure the on-going progress of people in economic growth terms. The feed-back to this is that with further economic growth, population growth rates will continue to fall.

Proponents of population controls and the doomsayer parables have not recognised the existence of the relationship between the institutions of property rights, prices and markets and the 'race' between population growth and resource availability. Does this indicate the existence of a 'little green lie'? Are these people trying to protect the rest of the world from what they see to be imminent disaster? Or are they protecting their own interests by lobbying to have governments enforce population growth controls on others? The choice is between benevolence for humanity that has ignored evidence that refutes the prospect of imminent disaster and an imposition of preferences by one group on another.

The imposition of preference argument is not one that necessarily involves the generation of monetary wealth. People calling for population reductions may be motivated by their own desires for a continuation of access to wilderness areas that few people visit. Their well-being is enhanced by access to space, peace and quiet. These are valid preferences for goods and services that may have public good characteristics that prevent them from being subject to the definition of property rights. So too may people be driven by the condition of environmental assets such as air and water. However, population control to achieve the supply of these goods and services is a very blunt policy instrument that may not succeed and may even trigger perverse outcomes. The appropriate action is for government policy to be developed with specific environmental targets being addressed by specific policy instruments.

In addition to specific policies targeted to achieve specific environmental goals, governments need to be mindful of their obligation to offer their citizens the greatest opportunities for them to make themselves better off. This obligation includes opportunities to make choices regarding family size that are most beneficial. The good news is that pro-growth policies that stress the importance of property rights and competitive markets provide those opportunities.

To the contrary, policies that limit the operation of competitive markets and reduce the incentives for people to act freely to pursue their best interests will have negative consequences in (at least) two dimensions. First, the ability of society to adapt to changing patterns of resource relative scarcity will be restricted. Second, people will find it more difficult to choose smaller family sizes.

5: Trade and the Environment

Proposition: Economic growth and trade are bad for the environment.

Economic growth, fuelled by international trade, means more pressure on scarce resources, including the environment. To protect the environment and to save resources for future generations, trade should be restricted to cut growth.

BUT

Trade and growth bring wealth to people. Wealth increases peoples' demands for environmental protection and the ability of society to provide environmental protection, especially through technological development.

Globalisation

Exchanging goods and services amongst people has always been a notable feature of human groupings. Whether those groupings were localised and predominantly family-based or more widely dispersed and involved only sporadic contact, people realised that there were mutual benefits available from trading.

However, over the last few centuries, two key innovations caused the world to 'shrink' and the capacity for people to 'group' to expand. The first was technological – the development of the chronometer – and the second was the development of a social institution – the acceptance of usury, the borrowing and lending of money. The chronometer enabled ships' captains to observe the longitude of their location, adding to their established capacity to determine their latitude from observing the sun's position. That single advance enabled long distance voyages in open waters to be feasible. The acceptance of usury enabled people to borrow the capital needed to 'float' long distance voyages. With such voyages possible from both technical and economic perspectives, trade between people from around the whole globe was given a significant boost. Voyages of discovery left the 'old world' in search of riches in the 'new world'. Loans were re-paid with the spoils of the expeditions, leaving profits to finance more ventures and the process accelerated until today when we see trade as ubiquitous.

Over time the process of trade has brought together the different peoples of the world. The 'global village' has arrived. People in Kenya may use mobile phones that were designed in the United Sates, manufactured in China with electronic components developed in Japan. Singaporeans may dine out on meals comprising Australian beef and Thai rice. An English family may drive a car that was built in South Korea. People is Scandinavia can eat mangoes in the dead of winter and the citizens of Costa Rica can enjoy a cool climate French wine.

The reason why trade is, and always has been, ubiquitous is that it makes people better off. People like to trade. If they don't want to trade, they are not forced to. So the Kenyans on their mobile phone are better off for having them. The Swedes indulging in their mangoes are happier as a result. But so too are the sellers of these products. The Australian cattle grazier profits from sales (they wouldn't be in the business if they didn't) as does the Thai rice farmer. Those people working in the Chinese factory where the phone was made are also better off: They receive wages that are sufficient to encourage them to leave what is otherwise their best earning position.

Because trade makes people better off, they seek to become involved. The incentive for self-betterment encourages trade to expand. People continue to search out ways of buying and selling that will make them better off. Sellers look to lower their costs of production, to improve their existing products and to develop new ones. Buyers, in their search for opportunities, stimulate sellers to action. Both buyers and sellers try to find ways to lower the costs of being engaged in transactions (searching, negotiating, contracting, financing, transporting) and other people (middlemen, traders, brokers, financiers, agents) provide innovative ways for them to do so.

The growth of trade has been a process that generated more trade. Because people could rely on trade to supply them with a wide variety of goods and services, they could concentrate more on the activity they did best, producing more of just one product and not having to produce some of everything they needed. Without trade, people needed to be 'self-sufficient'. With trade, they could specialise. Specialisation allowed people to do what they were best at and that meant the overall costs of production fell, thus saving scarce resources. It also allowed people to increase their incomes. With lower costs and more income, people were able to gain access to more goods and services that made them even better off. And because they were no longer 'self-sufficient' they had to become engaged in the trade network. The brokers and agents specialised in bringing buyers and sellers together by lowering transaction costs. Even though they produce no tangible goods, the services they provide are valuable because they allow more people access to the gains that are available from trade. People are willing to pay for those services because they get more value from them than they cost.

Why worry?

With such a good news story about trade, why are some people so vehemently against it? Why do we see protests outside conference venues where delegates to the World Trade Organisation seek to negotiate ways of reducing remaining barriers to trade such as tariffs (a tax on imports) and quotas (physical restrictions on the volume of imports)?

One very powerful explanation of this opposition is the fear that exposure to international competition will force some higher cost domestic producers to leave their industries. For instance, without protection from the competition offered by cheaper imports of agricultural products, some European farmers would be forced out of business and would need to find work elsewhere in the economy. That explanation is not the immediate focus of this topic, although it will become an important part of the 'little green lie' explanation at the end of the chapter.

The concern about trade that is the focus here is that of the environmental movement. This is a multi-dimensional concern. The first dimension is that with the growth in the volume of trade, more scarce natural resources are being extracted. This is the concern that economic growth, facilitated by trade, is causing the too-rapid exhaustion or depletion of the world's scarce natural resources. The second dimension is that with more trade and its associated increases in economic growth comes increased environmental harm. While this concern can be viewed as a component of the first – in that environmental assets such as air, water and biodiversity are scarce natural resources – it is split apart here as a separate element because it is often seen to be a separate concern from that about the rate of use of other resources. Finally, within the environmentalists' concern is an equity issue: The alleged environmental damage and resource over-exploitation is being primarily experienced in the developing world to the advantage of the developed world.

The concern regarding natural resource scarcity, too, is not the primary focus of this chapter. It has already been outlined in preceding chapters. It is a concern that amplifies the resource scarcity concerns of those who worry about the size of the world's population and is seen as a key reason why non-renewable energy sources are being run down too quickly and why consumption patterns need to be altered to ensure 'sustainability'.

What is of interest in this chapter is the environmental degradation that is argued to be accelerated by the process of growth. How this degradation is spread across nations is also considered.

Economic growth and environmental decay

Anyone who has visited the capital cities of the rapidly growing Asian economies will understand the environmental problems they face. The air at times is difficult to breathe and the tap water is often of dubious quality. The process of growth had an apparent adverse effect on the environment.

Burgeoning demand for power means more coal-fired power stations and more associated particulate and noxious gas emissions into the atmosphere. More industrial processing means more effluent going into rivers and streams. With economic development comes higher car-ownership rates and more smog as well as congestion. The greater consumption of all goods and services that comes with greater wealth places additional demands on the environment in its role as a 'sink' for the waste products of the processes involved in their production and consumption. The demand for goods and services in these rapidly growing economies comes from both domestic sources and overseas.

As well as the production and consumption processes creating environmental damage, the extraction of the natural resources used also has environmental consequences. Mining, forestry, agricultural and fishing activities all have the potential for biodiversity impacts and pollution.

The implication is that environmental conditions are negatively related to economic growth: As growth progresses, the environment deteriorates.

The actual picture is not quite so straightforward. Another way to look at the correlation between development and the environment is to examine the relationship on a country-by-country basis. Then what is apparent is an opposite relationship: the richer the country, the better is its environment. More developed countries have strict environmental policies in place to protect air and water condition and they have national park networks that effectively protect endangered species.

The 'Environmental Kuznets Curve'

The mixed picture emerging regarding the relationship between trade-induced growth and the environment is usefully reconciled by considering what happens to countries as they develop. This concept involves graphing the environmental condition of a nation against its level of economic development. The resultant curve – known as the Environmental Kuznets[1] Curve – was initially developed in the

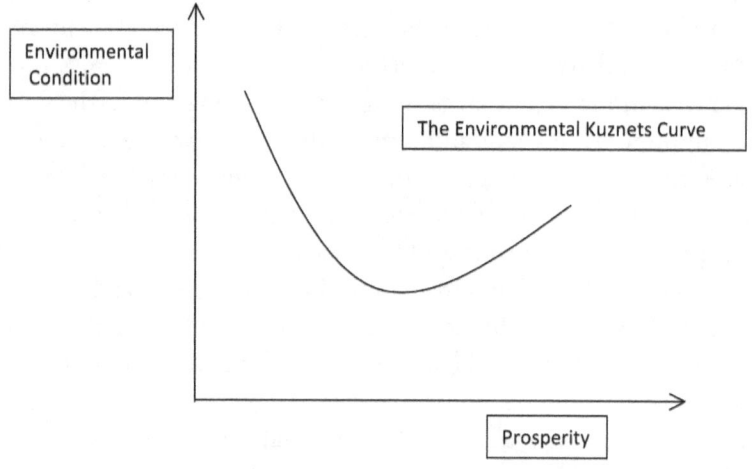

United States with just one environmental pollutant as its focus: sulphur dioxide, one of the key environmental pollutants resulting from coal-fired electricity generation. That relationship showed that, over time, and with growing levels of prosperity, even as the amount

---
1 Originally, Simon Kuznets developed the concept in relation to economic growth and income equity: as growth increases, income inequity in a society initially increases but eventually falls.

of coal-fired electricity being generated was increasing, the levels of sulphur dioxide emissions initially rose but eventually fell.[2]

It turns out that the Kuznets Curve relationship is useful for considering a wider array of environmental conditions. Typically, as a country experiences economic growth, there is a period in which environmental conditions deteriorate. However, as further development takes place and levels of prosperity build, the state of the environment tends to improve.

Of course, characterising 'environmental conditions' or the 'state of the environment' as a single metric is a dubious quantitative task. This was demonstrated in Chapter 3. There will necessarily be cases of specific characteristics or aspects of the environment in a country that will not always respond positively to growing levels of prosperity. For instance, irreversible environmental degradation such as species loss cannot be overcome, irrespective of wealth. Furthermore, over time, even reversible impacts may grow before measures are developed to control them.

However, evidence of improvement is amply illustrated by observations as diverse as water quality in the River Thames as it flows through the City of London[3] through to PM10 levels in the US.[4] From being the source of the cholera plagues that swept London in the middle of the 19th century, the River Thames now supports 350 species of invertebrates, 38 bird species and 300 plant species. PM10 refers to particulate matter in the air that is of 10 micrometres or less. It is a key indicator of the health impacts of air pollution. In the period 1990 to 2009, PM10 concentrations, averaged across 310 sites in the US, fell from 80.6 to 50.3 ug/m3,[5] with more than 90 per cent of sites recording levels lower than the nationally set standard (150 ug/m3) at all times within the period.

---

2 See Yandle et al. (2002) at http://www.perc.org/articles/article688.php
3 http://www.the-river-thames.co.uk/environ.htm
4 http://epa.gov/air/airtrends/pm.html#pmnat
5 Micrograms per cubic metre of air.

The recognition that environmental conditions in developed countries were improving, contrary to the impression that would be gleaned from exposure to the popular press and environmental lobby groups, stimulated Bjørn Lomborg to compile his book *The Skeptical Environmentalist*.[6] Its compilation of statistical series demonstrated that the environmental 'doomsayers' were at best exaggerating their case or, at worst, attempting to deliberately deceive by selective reporting. Needless to say, the publication of Lomborg's book met with considerable opposition, including a 'skeptical' contribution from *Scientific American*[7] that was rebutted by a piece in *The Economist*.[8]

Not all indicators of environmental conditions have shown improvement in all developed countries as they have become wealthier. For instance, the condition of Australia's inland rivers has deteriorated over the past few decades as a result of excessive extractions of water for irrigation.[9] Similarly fish stocks in the waters of European Union nations remain overexploited.[10]

What these apparently conflicting observations regarding growth and the environment reveal is that there are a number of processes at work when economies grow. First, there are the well-established economic principles of scarcity and choice: because our natural resources are scarce, society has to make a choice between 'extracting' them as the basis for economic growth and wealth creation and 'protecting' them. This implies that society can have trade, growth, and economic prosperity or an environment in sound condition. Such a trade-off between growth and the environment would be characteristic of the

---

6   Lomborg, B. (2001), *The Skeptical Environmentalist: Measuring the Real State of the World*. Cambridge University Press.
7   http://www.scientificamerican.com/article.cfm?id=skepticism-toward-the-ske
8   http://www.economist.com/node/965520
9   http://www.mdba.gov.au/programs/sustainable-rivers-audit
10 http://www.eea.europa.eu/data-and-maps/indicators/status-of-marine-fish-stocks/status-of-marine-fish-stocks-8

downward sloping section of the Environmental Kuznets Curve. For example, a forest can be set aside as a nature reserve or it can be harvested for its timber. Coal can be burnt to generate electricity to light and heat houses or it can be left in the ground without releasing any nitrous oxides into the atmosphere. Water can be diverted from a river to irrigate crops or it can be allowed to remain 'in-stream' to maintain wetland health. This is the classic case of 'you can't have your cake and eat it too'.

The second process is the one that allows for the Environmental Kuznets Curve to turn upwards. If additional natural resources can be discovered or their productivity increased, the potential is for improvements in both economic growth and environmental conditions to be achieved. The desirability of avoiding trade-offs and the achievement of 'win-win' outcomes is obvious. Everyone would prefer to have a situation where economic wealth is improving and there are no negative consequences in terms of environmental harm. Less clear are the processes that would initiate and sustain the improved availability and productivity of natural resources.

Combining capital

To understand the process whereby environmental conditions can improve as economic growth progresses, it is useful to consider the factors that are integral to the productivity of natural resources. The production of goods and services which is integral to trade and growth is necessarily a process of combining different forms of capital … natural capital, manufactured capital, human capital and social capital. More production can be achieved if the stocks of any of these forms of capital can be enhanced. For instance, if education levels can be enhanced so that the stock of human capital is increased, the production levels achievable given fixed natural capital can be increased – smarter workers and management plus the results of research and development are productivity enhancing. Additions to the stock of manufactured capital – higher technology machinery – applied to natural resources can increase productivity. Finally, improved levels of social capital – particularly strengthened institutional structures such as property rights – can enhance productivity. For instance, incentives for improved water productivity can be generated through the establishment of private property rights to water and the subsequent formation of a market.

In all cases, productivity enhancements from such improvements in capital stocks can lead to more economic wealth and better environmental conditions. For instance, human capital enhancements through research and development can be directed to more efficient methods of producing goods and services that prevent or even restore environmental damage. Reforming policies relating to the structure of society's decision-making processes (a form of social capital) can also improve economic wealth and environmental conditions. For instance, the establishment of well-defined and defended private property rights to agricultural land, fish stocks, and forests gives their owners incentives to improve their assets' productivity in both current and future uses. This gives owners the incentive to invest in soil erosion controls, reduce current catches to allow the fish to spawn, and replant harvested trees.

The important insight provided by the Kuznets Curve analysis is that higher levels of economic wealth allow these capital stock enhancements. Greater wealth means that more income can be spent on education and research endeavours to build human capital. Greater wealth allows more investment in manufactured capital: Cars with catalytic converters and coal-fired power stations with scrubbers on their stacks are more expensive than their more polluting alternatives. Greater wealth allows more income to be used for the establishment and maintenance of organisational structures such as the judiciary, police and the penal system to ensure property right defence and the avoidance of corruption. For instance, establishing systems of property titling that involve the accurate mapping of land or gauging of water flows are costly, just as patrol boats to enforce fishing quotas and wildlife sanctuary guards are expensive.

More demand for environmental care
Greater wealth is also important for ensuring that the environmental improvements – and the increases in capital stock that enable those improvements – are what society wants. Increased wealth means that people are less constrained in satisfying their wants. So with more wealth they are able to buy more of what they want. This is critical because unless changes are driven by increasing public demand for better environmental conditions, it is unlikely that they will be forthcoming. An important linking factor between wealth and demand for environmental improvement is knowledge – and its precursor, education. As wealth increases, there is generally an increase in society's demand for education and knowledge. Greater knowledge, better information flows, and an increased capacity within the general public to assimilate information bring with them an expanded public interest in the condition of the environment. Such awareness may relate to matters ranging from the personal health impacts arising from environmental degradation right through to concerns for endangered species.

With awareness – along with the financial means necessary to convert desire to action – comes demand for change that is expressed through both the private and public sectors. In the private sector, people become more interested in purchasing 'green' goods and services. People look for phosphate-free detergents and dolphin-safe tuna and sign up to be connected to renewable energy supplies. In the public sector of democratic nations, voters call upon their elected representatives to institute policies that protect environmental public goods and services. They vote for air and water pollution controls, creation or expansion of national parks, and regulations to help manage natural resources. Even in regimes where more totalitarian forms of government are in place, ignoring citizens' discontent about poor environmental conditions can be a de-stabilising factor. Furthermore, international pressures mean that there are 'spill-over' effects from the environmental demands emerging from other countries.

Policy implications

One policy temptation arising from the Environmental Kuznets Curve analysis is that there is no need for environmental policy: Because environmental conditions will eventually improve as an economy grows with more trade, all that is needed is a policy setting that encourages trade and thus growth. This is, however, a potentially dangerous conclusion to draw as it does not take into account the reasons why environmental conditions eventually improve.

One danger of a policy position based on a belief that 'growth will take care of the environment' is that in the period of a nation's growth when environmental conditions are deteriorating, some environmental thresholds may be breached. The clearest example of this would be species extinction. If the harvesting of timber or catching of fish caused species to become extinct, the environmental loss would be irreversible even if further growth stimulated the desire and capacity for environmental restoration. Other irreversible thresholds may exist in the pollution of waterways or soils. Without the ability to make good these adverse consequences, the costs to society of the environmental degradation would be on-going and likely to be substantial. Avoiding these costs would likely be a worthwhile goal for environmental policies – even in countries experiencing the early phases of economic growth.

A further problem with the 'growth only' policy logic is that while increased wealth can be argued to be a necessary condition for a reversal of environmental degradation trends associated with economic development, it is not sufficient. A close look at the reasons why environmental improvements take place with increasing levels of wealth is that the public puts increasing pressure on their elected representatives (or their otherwise empowered leaders) to install increasingly demanding environmental policies. Because so many environmental goods and services involve public good characteristics that prove difficult for markets to supply, public policy is a key element of an improving environment. This can only occur when governments are not only sufficiently well financed to be able to act, but also when there is sufficient public support to provide a mandate for action. In the policy context, it is therefore important for economic growth to be a goal of government but that specific policy strategies to facilitate environmental protection are also developed. This is consistent with the Tinbergen Principle mentioned in Chapter 2, of adopting policy measures equal to the number of policy challenges: individual policy measures are unlikely to be perfectly synchronised to achieving multiple goals and thus likely to deliver conflicting outcomes across multiple goals.

Contemplating the converse situation reinforces this conclusion. This involves accepting – for the purposes of the thought experiment – that economic growth is associated with environmental decay and that to correct for that decay, growth should be restricted. To do this, governments might introduce limitations or even bans on domestic and international trade. Certainly the outcome would be reduced economic well-being, but would there be an associated improvement in environmental conditions? For a number of reasons, the answer is most likely 'no'.

With less wealth, both the capacity to develop environment-improving capital stocks and the public demand for such action would be diminished. People struggling to meet basic needs of food and shelter are more likely to think in the very short term than act to protect assets that will yield long-term returns such as the environment. However another important aspect of environmental harm comes with the limits placed on people's ability to specialise in activities for which they have a comparative advantage. Specialisation allows increased efficiency of resource use. If goods cannot be imported from specialist producers, less efficient domestic suppliers will be relied upon for supplies. They will use more scarce resources in their production activities, and potentially, cause more environmental harm than would be generated in a location better suited to the production process. An example is in agricultural production. Banning imports from a country having a comparative advantage in open-range, extensive livestock production may mean that local people have to buy locally produced meat that is raised intensively at higher financial cost and with greater environmental impacts through nitrate contamination of water sources.

The consequences, therefore, of trade restrictions are not just reduced economic wealth but also more environmental harm. Attempting to protect the environment by restricting growth would likely have the opposite outcomes to those hoped for by environmentalists.

Relocating environmental harm

One argument used in attempting to refute the logic of the Environmental Kuznets Curve is that rich countries can protect their environments by shifting environmentally damaging production processes to less developed countries. Hence, it could be argued that the reason why PM10 concentrations in the US have declined as economic growth has proceeded is that goods and services involving the production of PM10 emissions are now being imported from heavily polluted developing countries such as China and India.

Certainly the consequences of international trade involve the constant revision of comparative advantages. The relative costs of producing goods and services change as the relative scarcities of all the different resources involved change. Changing levels of wealth alter relative wage rates, and technological change impacts on production methods and transportation options as well as the mix of goods and services that people want.

These changes mean that the patterns of production shift globally. Regions where steel-making was once the dominant industry, such as around the Great Lakes in the United States, are now referred to as the 'rust belt'. China, Taiwan and Korea have newly established steel-making centres. Dairy farming in Europe is in decline (despite European Union subsidies) as the industry booms in New Zealand (without government assistance).

The same changes in relative costs across the world also impact on the comparative advantages of countries in terms of their capacities to process and assimilate environmental wastes. For instance, mineral ore processing is likely to be less environmentally damaging (that is, costly) if carried out in remote locations with sparse populations and robust ecologies than in the midst of a heavily populated area. Hence, the relocation of aluminium smelting operations from Japan to Australia in the 1970s ensured lower operational costs and lower environmental costs. So while the relocation of industry doesn't change the amount of emissions produced, it can lower the costs of those emissions because fewer people are affected by them.

So as the populations of countries become wealthier and more environmentally conscious, the costs of environmental damage rise and comparative advantages shift. This is not just a phenomenon experienced at the international level. It can also happen within a country. For instance, industries with high pollution profiles have historically relocated out of urban areas to sites that are better able to assimilate wastes. In other words, the costs of waste disposal are relatively lower. The tanning industry once located in the industrial suburbs of Sydney is a good example. During the 1970s this industry relocated to surrounding regional areas where the wastes of the production process could be disposed of over a wider geographical area to take advantage of natural assimilation processes. A driving force of the relocation was the imposition of more stringent environmental waste regulations in the city area, reflecting the community's increased sensitivity to the condition of the local waterways.

On a global scale, the relocation of industries to take advantage of comparative advantages results in lower overall costs, both in terms of production and environmental damage. Restricting trade so as to prevent relocations has the consequences of increasing costs, with the implication that more scarce resources are used than otherwise needs to be the case.

One such type of restriction that impacts on the ability of nations to assert their environmental comparative advantages operates through the Basel Convention on the Control of Transboundary Movements of Hazardous Wastes and Their Disposal,[11] otherwise referred to as the Basel Convention. Under this treaty, which has 175 parties and came into effect in 1992, the movement of hazardous wastes between nations is restricted. The Convention is particularly focused on preventing the transfer of hazardous waste from developed to less developed countries. Amongst the prominent materials on the list of hazardous wastes covered by the Convention are biomedical and healthcare wastes, used oils, used lead acid batteries, Persistent Organic Pollutants (POPs), pesticides, Polychlorinated Biphenyls (PCBs), electronic and electrical waste ('e-waste') such as mobile phones and computers, ships destined for dismantling, mercury, and asbestos. The treaty restricts movements of these wastes even when the object of the movement is to take advantage of the comparative advantage that developing countries may have in the recycling of products. For instance, many recycling operations, such as the dismantling of electronic componentry and the recovery of precious metals, are highly labour intensive. The comparative advantage in such recycling operations that is offered by developing countries through their supply of low-cost labour is forgone because of the Basel Convention restrictions. The dismantling of ships on India's east coast is similarly adversely affected.

By preventing trade in wastes, the Basel Convention prevents the achievement of cost savings through comparative advantage.

---

11 http://www.basel.int/index.html

Furthermore, products that otherwise could be recycled are now disposed of in developed countries either without being recycled at all or recycled at far higher cost (and hence greater use of scarce resources) than could be achieved through trade. In addition, the livelihoods of those in developing countries who were profitably engaged in recycling activities are lost.

Certainly the Basel Convention has prevented the occurrence of unfortunate incidents such as those that stimulated its formation. These included the Khian Sea incident when a barge loaded with toxic ash from an incinerator in Philadelphia spent 16 months in 1986–87 searching for a location to dump its cargo and the 1988 Koko case when 8,000 barrels of hazardous waste from Italy were shipped to Nigeria where it was stored on farmland for $100 a month rent. However, the unintended consequences of the Convention in terms of higher costs, lost livelihoods and even greater environmental harm have been profound. The question raised again is the suitability of a restriction on trade as a means of dealing with an environmental issue. A much better approach is to target the environmentally damaging activities with policy instruments that are specific to the case. Such instruments could be enacted by developing country governments. If the necessary institutional structures are not sufficiently well formed to allow that to be effective, the responsibility could be accepted by the governments of the developed countries where the wastes originate.

A 'little green lie'?

If trade and the economic growth and wealth it brings are, in the long term, associated with environmental improvements and if restrictions to trade are harmful in terms of both people's wealth and the condition of the environment, why do we see environmental groups proposing trade restrictions? It would seem that trade and growth create both economic and environmental advantage. Why try to 'kill the goose that laid the golden egg'?

A big part of the answer to that question arises because not everyone is advantaged during the process of growth. With growth comes change. Comparative advantages shift not just for nations but for individuals as well. The skill sets developed by steel workers in Michigan may not be as attractive to employers in 2011 as they were 50 years ago. Likewise, the skill sets of Chinese farmers may not be as well rewarded in Shanghai as if they were trained as metallurgical engineers.

Hence, with changing economic structures brought about because of trade, some people within each nation's population will find their capacity to earn threatened in some way. In other words, even though, overall, a nation may be better off with trade and growth, there will be individual winners and losers within its population. The losers – including firms and the employees of those firms that are losing their comparative advantage – are likely to resist the change. A likely strategy to resist change is for the losers and their representative organisations, including industry lobbies, and the trade unions, to generate political support for anti-trade measures such as tariffs and quotas on imported products. Outright bans may be sought on the importation of some products that can be argued to breach health or quarantine standards. If trade restrictions are implemented, these groups will avoid the pressures imposed by change and be better off, at least in the short term.

In instances where these industry and labour groups can find support from the environmental movement, their case will be so much the stronger in the political arena. So the concerns of environmentalists that trade brings about environmentally harmful economic growth present an opportunity to form an alliance of interests. Protests designed to disrupt the meetings of the World Trade Organisation which attempt to negotiate reductions in trade barriers are therefore notable for the composition of the crowd. Placards displayed show allegiances to unions, green groups and industry councils. With such a broad-reaching array of opposition, it is perhaps not surprising to see how politically difficult free trade negotiations are. Entrenched interest groups oppose the changes that will arise from freer trade. The negative effects of change are concentrated on these groups and they are readily able to mobilise political interest in their cause. In contrast, the beneficiaries of freer trade are more difficult to identify. They are the developers of new businesses and their future employees and the consumers who benefit from a greater diversity of products at lower prices. Similarly, the environmental effects of trade restrictions and the beneficiaries of environmental improvements enjoyed in the fullness of time as growth progresses are complex to disentangle and do not involve immediate and transparent gain.

So the story of the 'Baptist and the bootlegger' once again emerges as an adjunct to the 'little green lie': Vested interest groups with a lot to lose from change align themselves with environmentalists who are focused on what are the apparent negative consequences of trade on natural resource use, including pollution. Propagating the 'economic-growth-environmental-decay' 'little green lie' helps to support the joint cause against trade.

The irony of this position is that, in the longer term, trade restrictions are bad for both parties. For industry and labour, barriers to trade cause economic stagnation that means fewer opportunities for wealth creation for others and themselves in the longer term. For the environmentalists, a stagnating or shrinking economy means less interest in the environment and less wealth to be used in investing in the environment. Restrictions to trade can even have direct negative impacts on the environment because of the inability of nations and people to pursue their comparative advantages in production processes and in dealing with environmental threats. More scarce natural resources will be used less efficiently in producing goods and services and in waste disposal.

Importantly, the rejection of the trade and environment 'little green lie' does not mean that an auto-correction mechanism is in place that will take care of the environment. What it does suggest, however, is that environmental policy needs to be developed on a case-by-case basis, rather than relying on blanket instruments such as a ban on foreign trade. In each case the rationale for a policy needs to be well-established and a process developed to ensure that a proposed policy does make an improvement to the well-being of society and that it makes the best improvement of any of the alternatively available policies. In most circumstances, the appropriate tool for such a vetting process is benefit-cost analysis. For example, consider the case where there are concerns that trade is causing an excessive rate of timber harvest from a nation's forests. The policy analysis for that case should involve the development of a rationale for action followed by a benefit-cost assessment. The benefit-cost analysis should compare the proposed policy's impacts relative to the do-nothing new option and other intervention strategies. It can never be assumed that just because a problem is perceived to exist that it will improve the well-being of society to do anything different. The problem may be one only for a small group within society or it may be that any policy aimed at addressing the problem is incapable of achieving anything better than the current situation.

Important in any analysis of policy options is the identification of unintended consequences. Policies may be well-intentioned and have an impact on one dimension of the problem being addressed; but they may also cause perverse outcomes on another dimension that are ultimately more of a problem. The case of the Basel Convention and its negative impacts on recycling resources is a good example. Most environmentalists would argue strongly for recycling, yet the Basel convention, designed to promote environmental management, has had the effect of diminishing the extent of recycling.

6: Waste

Proposition: No waste should go to landfill.

Waste should not be 'wasted'. It is a resource that can be re-used and re-cycled. Sending waste to landfill means that more 'virgin' resources must be harvested/mined. Waste in landfill can also be a source of air and water pollution.

BUT

Recycling and re-using 'waste' is a process that uses scarce resources. Policies that prevent landfill disposal can cause more resources to be used than they save and do not necessarily reduce virgin resource use. Landfills need not be pollution sources.

What a waste!
In 2009, landfills in New York State accepted a total of 7.6 million tons of household, commercial and institutional solid waste materials.[1] In addition, incinerators in the state processed approximately 3.9 million tons of solid waste.[2] These volumes of material, mirrored in cities and states around the world, create a range of management issues for local authorities ranging from the logistics of collection through to the siting of disposal areas. But they also create concerns for people who firstly worry about the continued ability of the world to support the extent of resource use that generates the waste. Secondly, they

---
1 http://www.dec.ny.gov/chemical/23682.html
2 http://www.dec.ny.gov/chemical/23683.html

worry about the lost potential for rubbish to be re-used or recycled. And there is also concern about the damage done to the environment from the waste dump itself.

The volumes of waste going to landfill are a visible indicator of the amount of resources being used by society. The burial of waste in landfills is also a clear signal that the resources once used to create the goods and services that have become waste are no longer available for any future use.[3] They represent a form of non-renewability in resources. There is a sense of loss, or perhaps frustration, in that waste sent to landfill constitutes resources that will never be productive again. It represents the second law of thermodynamics at work: Both energy and matter are continually becoming less useful.[4] Finally, waste dumps are rarely environmentally pleasant places. They can be sources of unpleasant odours, dust and emissions (transmitted beyond the site by both water and air) of varying degrees of toxicity depending on the materials being dumped at the site. Waste dumps are a common cause for the 'NIMBY' or 'not in my back yard' syndrome, reflecting the unwillingness of local residents to accept the locating of waste dumps in their neighbourhoods.

These concerns have been primary in the motivations of those such as the Zero Waste International Alliance[5] and the US-based Zero Waste Alliance[6] who have sought to influence the behaviour of individuals, corporations and public policy-makers who deal with waste. The mantra 'reduce, re-use and re-cycle' has been coined as part of the 'waste hierarchy'[7] concept to convince the public, corporations and government alike of the virtues of not sending material to landfill. This

---

3   Although the mining of old landfill sites to retrieve former wastes that have now become sufficiently valuable to warrant recovery does take place: http://wasteage.com/mag/waste_exploring_economics_mining/
4   http://www.physicalgeography.net/fundamentals/6e.html
5   http://zwia.org/joomla/
6   http://www.zerowaste.org/index.htm
7   http://www.wastenet.net.au/information/hierarchy

hierarchy is often expressed graphically as a pyramid with reducing waste achieving the high status of the peak of the pyramid and disposal, being considered only as a last resort option, sitting at the base of the hierarchy pyramid.

The ultimate expression of the waste hierarchy has been the adoption of the 'no waste to landfill' goal by a number of governments around the world. For instance the Scottish Government supports Zero Waste Scotland.[8] The city of San Francisco aims to have zero waste by 2020[9] and a number of states in Australia also aspire to a zero waste outcome[10] including, until recently, the Australian Capital Territory that revoked its policy of 'no waste by 2010' in the light of an upward trend in the amount of waste being sent to landfill.[11]

What is waste?

To give consideration to these concerns and the policies that have arisen because of them, the first step is to define the term 'waste'. That is not a simple task in itself.

At a superficial level, the task may seem easy. New York State[12] defines solid wastes as 'any discarded (abandoned or considered waste-like) materials' but recognises that the 'fine print' definition includes 'garbage, refuse, sludge from a wastewater treatment plant, water supply treatment plant, or air pollution control facility and other discarded materials including solid, liquid, semi-solid, or contained gaseous material, resulting from industrial, commercial, mining and agricultural operations, and from community activities'.

However, that style of definition does not allow for a consideration of why a resource which has value in use becomes a waste that is sent to

---

8   http://www.zerowastescotland.org.uk/
9   http://www.sfenvironment.org/our_programs/overview.html?ssi=3
10 http://www.zerowaste.sa.gov.au/
11 http://www.abc.net.au/news/video/2010/02/26/2831779.htm
12 http://www.dec.ny.gov/chemical/8732.html

landfill. This is where the definition has to move from being material-based to one that is based on the choices that people and businesses make and hence one that uses economic principles.

The definition of waste is a tricky one because what one person considers a waste that needs disposal is another person's bounty. Observing the person rummaging through the garbage bins of affluent cities or the 'rag-pickers' in the slums of Delhi demonstrates this, but so too does the use of used tyres in road base and the sale of steam from a power station for heating homes.

In thinking through what to do with a resource they own, people will weigh up the value they can achieve from it in a range of alternative options. One option is for the resource owner to use the resource themselves to make themselves better off. This will entail bearing some costs but enjoying the benefits of use as well. Another is the sale of the resource to an alternative user. Finally, they can choose to dispose of it as 'waste'.

Each of these options will be weighed up against the others in terms of how they impact on the decision-maker. For instance, the net benefit enjoyed by using the resource themselves (either in consumption or in the production of goods for sale) may be greater than the return that could be enjoyed from selling the resource or the costs of disposing of it. Then the choice is most likely to be to hold on to the resource. But what if no-one wants to buy the resource and keeping the resource will only generate a loss? In that case, so long as the loss from holding on to the resource is greater than the costs of disposal, then the resource will become waste. Furthermore, even if resource owners cannot achieve any net benefit from their own use of the resource, it may not be 'waste'. Another person may have different skills or interests that allow them to use the resource profitably.

For this reason, a wrecked car that is of no more use to its owner is not necessarily sent to landfill as is. It may become the source of spare parts and its steel melted down and reprocessed for further uses.

Likewise, a ship that can no longer be operated profitably by its owner will not necessarily be taken out to sea and sunk. It may be beached in India and cut up for its steel to be reprocessed. Off-cuts from a clothing manufacturer may not be burnt but may be sold for use as auto mechanics' cleaning rags. Bark stripped from trees in the processing of sawn timber may not be dumped but rather be sold as landscaping material. Even the solid waste incinerated in New York State was the source of around 2 million megawatt hours of electricity in 2009. In a manufacturing process, what may be a waste in one operation may be a joint product in another.

It is only if there are no buyers for a resource that is unwanted by its owner (because it cannot generate a positive net benefit) that the disposal option will be considered. This would indicate that there are no potential uses for the resource that can make society better off through the creation of a net benefit. Even then, the cost of disposing of a resource provides an incentive for its owner to consider storing the resource in the prospect of profitable uses being developed over time. The storage costs net of the expected future returns have to be less than the disposal costs to make that worthwhile.[13]

Why try to reduce waste?
This definition of waste suggests that the creation of waste is not something that is taken lightly by resource owners. If a resource has a value either to its owner or to someone else, it is unlikely to be sent to landfill. And because disposal itself involves costs, that option is even less attractive. People do not generate waste specifically to enjoy a gain. They create waste as an unfortunate 'by-product' either of production or consumption processes that yield them net benefits.

---

13 Note the significance of the word 'owner'. It is only with the rights to a resource being well defined and defended – so that the resource is effectively 'owned' – that these incentives for resource use come into play. Without those rights, there is no incentive to seek out those who might value a resource.

Wastes are unfortunate because they are costly in their disposal.

Given this insight, what is the rationale for policy interventions that seek to reduce waste either by encouraging reduced resource use, re-use or re-cycling, or by restricting the disposal of waste? If people don't want to create waste in the first place, why is it necessary to make it even less attractive to create waste?

The first motive of those who seek reduced waste levels comes from the concern for the overall rate at which resources are being used. The call for reduced waste is frequently associated with claims that rates of consumption are 'unsustainable': If people didn't consume so much, there would be fewer scarce resources used and hence less waste created.

The implication of this concern is that resources are 'too cheap' and so we use too many of them, too quickly. This would mean that the resource scarcity signalling function of prices has somehow become distorted so that resources are too cheap. If this is the case, it is important to understand why the market is not delivering accurate signals. If they are not, it is an argument in favour of addressing the reasons why they are not working well. Prime candidates would include the lack of well-defined and well-defended property rights over the resources that are alleged to be over-exploited. Seeking to achieve changes in resource use patterns by controlling waste flows is a very blunt policy instrument to address these fundamental causal factors. It is unlikely to deliver outcomes that are in society's best interests because it is so poorly targeted.

The second motive is the lost potential of the resources that are sent to landfill. The assertion is that humanity should mimic nature 'in which nothing is wasted'. For example, the Zero Waste Alliance[14] states that it is 'Following nature's model ... Working for the elimination of waste and toxics'. The implication of this motive is that there are socially beneficial uses for the resources currently being sent to landfill.

---

14 http://www.zerowaste.org/index.htm

However, for some reason those use opportunities are not being taken. This situation would occur in a number of circumstances. Market prices for the goods and services created from the use of the resources may be 'too low'. The costs of their use may be 'too high'. Alternatively, the costs of their disposal to landfill may be 'too low'. The expressions 'too high' and 'too low' are used here to reflect the situation where market prices are somehow not reflecting the true picture of relative resource scarcity. Again, whether or not prices are working well is a matter for investigation on a case-by-case basis. For instance, costs for disposal at landfills may be 'too low' because the environmental impacts of landfill sites are not fully integrated into those costs. This relates to the final motivation: Concern about the environmental condition of waste disposal areas.

Waste by its very nature consists of things people don't want. It is therefore not surprising that people don't want to have their waste and the waste of others around them: Hence the dislike felt by people for having waste disposal areas located in their local region. Landfills and less formal disposal areas can be associated with odours, traffic congestion, noise, vermin, leachates and wind-borne waste. These are impacts on resources that do not have well-defined or defended property rights and so are not necessarily traded in markets. The result is that the prices charged for the use of a formal disposal area may not reflect full resource scarcity as those bearing the full range of costs are not paid for that imposition.

With costs of disposal to landfill in such circumstances thus being 'too low', 'too much' waste may be created and 'too many' environmental impacts endured by local residents.

Particularly in developing countries, there are issues associated with waste disposal outside of formally defined landfill areas. And in developed countries, 'litter' and illegal dumping of waste are sources of concern. This sort of problem can be defined in terms of breaches of property rights. Where wastes are dumped on the property of others (including common and state property) the rights to that land (or water) are being breached. This can occur because of inadequate enforcement of rights: when wastes are dumped the people responsible are not punished for their breach. However, the problem can also be defined in terms of a complete lack of property rights. In other words, if no-one owns the resource into which the waste is dumped then there is 'open-access'. In both cases, the costs of disposing of the waste (in terms of the environmental impacts) are not reflected at all in a price, as there is no price charged. Waste dumping in that case becomes a very low-cost option for those seeking to get rid of their unwanted resources. Again, the result is 'too much' waste and 'too many' environmental impacts.

Even in these circumstances, policies designed to reduce the amount of waste being created are blunt instruments. The problem of excessive rates of waste creation and the environmental impacts of waste dumps can be caused by factors relating to the management of landfill sites (such as the price being charged to dump waste). The appropriate action to take should therefore be directed at ensuring that the environmental costs are borne by those creating them. This would involve payments being made to those living in the vicinity of landfill sites. This in turn would make living adjacent to a landfill a more attractive proposition. The payments made to locals would also increase the costs of running the landfill sites. Passing on these costs to users would decrease the incentive to create waste.

Furthermore, the proliferation of waste disposal outside formal landfill sites can be caused by poorly defined or defended rights to land and water where dumping occurs. Vesting rights to the affected resources and then ensuring the prosecution of offences against those rights are the first steps to be taken. Bigger fines for dumping outside of designated landfill sites and increased probabilities of being caught mean that the expected costs of illegal dumping are increased. People will then be more likely to use formal landfill sites as they become a relatively less expensive option.

It is worth noting that, frequently, the waste management issue requires both of these directions to be pursued. If environmental costs are embedded into the prices charged for formal landfill disposal, those higher costs give rise to the incentive to dump illegally. This incentive has to be managed by increased vigilance in the definition and defence of resource property rights.

Waste reduction policies
Another way of looking at the policies that zero waste proponents advocate is to analyse their likely impacts on the well-being of society.

One such set of policies involves the encouragement of recycling activities. Such policies include the imposition of recycling requirements (such as compulsory deposits on beverage containers) and the subsidisation of recycling activities (such as the low-cost provision by local governments of separate garbage bins for recyclable material).

Numerous reasons are given for such policies. For instance, in the case of paper recycling, concern is given to the rate of harvesting forests for paper pulp. For aluminium cans, energy savings in the processing of bauxite through to aluminium are used as justification. This means more energy resources are left by the current generation for use by future generations. Environmental damage avoided at mine sites is also claimed as an advantage. Litter reductions are another source of rationale, especially for glass beverage containers which when broken in public places pose a health threat.

Does recycling achieve its stated goals?

Take, for example, the case of a policy that seeks to encourage paper recycling which has as its goal a reduction in the logging of forests. Given that the recycled product is a good substitute for the paper made from virgin materials, the increase in the supply of paper caused by the recycling incentives is likely to have the effect of reducing its price. This could have the unintended consequence of increasing the demand for both recycled and virgin paper (depending on their relative costs of production, including any subsidies). This then could increase the harvesting of forests. In other words, the regulation to encourage recycling causes a distortion to the resource scarcity signal that the price of paper provides.

If the demand for paper responds to the increased amount of paper available simply by expanding[15] – and there is no price fall – then the recycling assistance will have no effect on the amount of virgin material harvested.

A second example is the power and bauxite savings resulting from aluminium recycling. If there is 'crowding out' of the virgin material by the recycled alternative (as unlikely as that may be) then it is true that there would be more resources passed on to users in the future. But what is the value of those resources to future users?

Two factors are important in answering that question. First there is the change in price of the resource that could be expected to occur through time. This may be a price rise if its (non-renewable) stocks are run down over time and immediate substitutes do not become available. Or it may be a price fall if new deposits are found or some technological change results in the development of cheaper alternatives.

The second factor is just how long into the future will the resource savings be enjoyed? This brings to the fore the issue of time discounting. The evidence from peoples' choices is that they like to have things now rather than in the future. They engage in actions like paying to borrow from others (in the form of interest) so that they can have what they want now. For people to give up having what they want now and postponing it to the future (that is, they save) they require payment (in

---

15 Technically, this is the case of an 'elastic' demand for paper. Where demand is perfectly 'inelastic' – that is, where the supply shift causes no increase in demand but the price drops – then recycled product will displace virgin materials if they have lower costs (including the subsidies paid to producers). Where there is a range of substitutes for a product (such as electronic copy for paper copy or plastics boxes for cardboard ones), demand is likely to be more elastic.

the form of interest). The implication of these revealed preferences for 'time' is that the value of receiving good things a long way into the future is less than what it would be if they were received now. For example, $110 in one year's time is equivalent to $100 now, at an interest rate of 10 per cent. That process – called 'discounting' - means that $100 to be received in 50 years' time would only be worth 85 cents[16] now.

The implication of these two factors is that the value of any virgin resources saved by recycling may not be very much at all. For a start, if there is no expected price change for a resource over time, the discounting process using market interest rates (correcting for inflation) will render the value of future use of the resource smaller the longer into the future the stocks of the virgin resource last. For relatively abundant resources like bauxite and coal, this could be a long time. To counter the impact of the discounting process, the prices of the resource would need to be increasing significantly over time.[17] Experience to date has been predominantly of falling resource prices (corrected for inflation). If this is maintained – as would be predicted given new discoveries both of resource reserves and substitutes – the value of 'saved' virgin material would be even lower, if indeed there has been any savings at all.

The question then is: What is being lost in order to enjoy this (albeit low) benefit from saving resources? The reason recycling needs to be encouraged by government policy indicates that recycled material is more expensive than the virgin alternative. What is being lost therefore is the use of the cheaper virgin material. Put another way, the cost to society as a whole of achieving the benefits of increased future use of resources is the premium that has to be paid for recycling.

---

16 The process of discounting is the opposite of compounding. Under compounding, a dollar put aside now is worth $(1+i)n$ in n years' time, at a discount rate of i%. The (discounted) value of a dollar received in n years' time is $1/(1+i)n$.
17 The Hotelling rule detailed in Chapter 1 requires prices to rise at the interest rate to reflect rising relative scarcity. This would keep the present value of prices constant.

It is important in this context to recognise that recycling is not a costless activity. It is not costless because it involves the use of scarce resources. Those resources are not the virgin materials that are the focus of the recycling regulation (for instance, the forests in the recycled paper or the bauxite in the aluminium example). They are resources like the energy used to collect and process the used paper and old aluminium cans, the chemicals used in the processing operation, the storage facilities used to house the stock of collected cans and paper etc. All of these resources have positive prices indicating that they are scarce and also have alternative uses.

Recycling therefore involves the substitution of one set of scarce resources for another. Which of these alternative sets of resources are the scarcer and so which of them should be saved? The usual way society works out the answer to this question is with reference to prices: Which set of resources costs less? The need to impose regulations to stimulate recycling would indicate that it is not the cheapest (least resource-using) option. Hence, the cost premium on recycled products which without government policies would not be produced.

The danger that is posed by policies that encourage recycling is that they force society into using resources that are scarcer than the ones for which protection is being sought.

Again, if it is the concern that forests are being harvested 'too quickly' because the price being charged for the use of the forest resource is too low, then recycling is a blunt instrument to ensure the protection of the forest. A more appropriate way to ensure the protection of the forest would be to investigate the property rights structure covering the forest and addressing the inadequacies in that respect with more direct measures. For instance, timber from forests owned and managed by the state may be currently priced for the benefit of businesses reliant on the forest asset for their production inputs. Ensuring competitive tendering for access to the resource would be an effective way of sending a price signal that reflects real levels of forest resource scarcity.

If environmental damage from mining bauxite is the concern, then again, a range of policy approaches is available. In developed countries, these may involve the enactment and enforcement of environmental damage liability laws reflecting existing property rights. So the owners of properties adversely affected by contaminated water run-off from the mine site could claim damages from the mine owner. Where legal and administrative structures are less robust, as in less developed countries, those interests concerned about environmental damage would need to mobilise their concerns to pay for the implementation of improved practices.

These are actions that are specifically directed at achieving improvements in the well-being of people at the source of the problem. By targeting waste as the policy vehicle, the danger is that the desired results will not be achieved and people will be made worse off in the process through the higher costs associated with waste recycling policies.

Landfills

Policies designed to reduce the amount of waste going to landfill also give no guarantees that the environmental problems associated with landfills – particularly as they affect nearby residents – will be any less. The same problems associated with a landfill site of two hectares can be created by a facility that is half that size. While the extent of waste may be a factor in determining the offsite impacts of a landfill site, the way in which it is managed is also an important determinant. A relatively small landfill site that is poorly managed can be the source of more environmental impacts than a well-managed large site. The technology of landfill management has developed to the point where it is possible to ensure the almost complete isolation of wastes from the environment. The lining of pits with clays and plastics[18] and their

---

18 http://glossary.eea.europa.eu/terminology/concept_html?term=landfill%20base%20 sealing

subsequent capping when filled can prevent offsite contamination and then provide a source of methane gas for energy generation as the waste material decomposes over time.[19]

Of course, the technology involved in sealing landfill sites is costly, implying the use of scarce resources. These costs need to be weighed up against the benefits achieved through managing landfill sites in this way. As people become wealthier and the environment features more in their preferences, it is likely that they are more willing to pay the costs of sealing landfills. This is the experience in developed countries, especially where land is expensive and the use of former landfill sites for development projects such as sporting facilities creates significant social value: the costs are warranted by the benefits. It is generally not the case in poorer countries.

Managing solid waste in developing countries is a vexed issue. It is possible to design and manage landfill sites to vastly reduce their environmental impacts, but that is expensive. To fund these costs, the fees for use would need to be high. With higher user fees, there is less likelihood of the facility being used and a greater probability of illegal dumping and simple littering. And to prevent illegal dumping and littering requires resources to be devoted to enactment and enforcement of property rights. These too involve costs that may not be easily afforded in the developing country context.

This conundrum is not only experienced in developing countries. Wealthy countries also find that increasing landfill access charges can induce people to dispose of their wastes in other ways, even if they are illegal.

Informal solutions

One strategy used widely to try to overcome this complexity is to try to encourage the formation of social attitudes that reject littering and

---

19 http://www.thinkgreen.com/landfill-gas-to-energy

dumping. Such attitudes are a form of social rule that is established at an informal level. They mean that people are less willing to engage in littering and more willing to put pressure on others not to litter. They also mean that formal actions taken to prosecute those who are caught dumping illegally are widely supported in the community. The 'naming and shaming' of offenders can be a powerful incentive against littering when strong community attitudes prevail.

The informal rules embedded in social attitudes are alternatives to the more formal rules that society can institute to prevent dumping and littering such as regulations and even the criminal code. The informal rules are clearly less costly in action than the formal rules. Once established, they are internally enforced by the actions of individuals without external coercion. However, they can take a long time to establish if the current social 'norm' is to accept littering and waste dumping. They can also be costly to establish in terms of resources devoted to advertising campaigns over extended periods.

Educational programmes about littering conducted in schools, for instance, can influence children's attitudes that in due course shape future community attitudes. Community action can also help. For instance, in Australia, campaigns such as 'Clean up Australia'[20] and 'Keep Australia Beautiful'[21] including its 'Tidy Towns' competition are promoted widely through local community events and Australia-wide advertising campaigns. These are community initiatives that are supported by government funds on the basis of the public good elements of the benefits they provide for society.

So for developing countries an important starting point is the establishment of educational programmes that shift the social norm from accepting littering to rejecting it. Such programmes, along with broader community-based advertising campaigns, are likely to be more successful as levels of wealth increase: a cleaner environment takes on

---

20 http://www.cleanup.org.au/au/
21 http://www.kab.org.au/

greater importance when more basic needs have been satisfied. They also need to be designed to ensure that their costs are commensurate with the community benefits associated with the decreased incidence of littering they generate.

Collecting waste

However, even with strong social norms in place, the relatively low costs of rubbish disposal outside of landfills can be a strong driving force. For this reason, many authorities around the world have instigated practices that effectively subsidise formal waste disposal. An example is the charging of a fixed annual cost for household rubbish collection: Households are charged a fixed fee each year to have a rubbish bin collected each week. The implication of this type of scheme is that up to the threshold level of the capacity of the rubbish bin, extra waste costs nothing more to collect. While this type of scheme gives no cost disincentive to create waste, at least up to the threshold level, it does make formal waste disposal a more attractive option relative to illegal dumping or littering.[22]

Put simply, societies appear to be willing to accept a somewhat larger amount of waste being created in order to have less of that waste being disposed of as litter or by illegal dumping. The costs of more waste and higher landfill management costs are traded off against the benefits of less litter/dumping and lower costs of policing such illegal acts.

So for developing countries, as well as educational campaigns, provision of low-cost landfill disposal options and household collection systems is a step toward cleaning up unsightly and unsanitary street waste. Making sure the provision of these services is carried out on a competitive basis by the letting of contracts through open tendering is a necessary part of ensuring the best use of available resources.

---

22 The same is true for the provision of 'free' public toilets as a means of reducing the health risks and general disamenity associated with people 'relieving' themselves in public spaces.

One side-effect of policies that make landfill more competitive than illegal waste disposal is that more landfill sites need to be developed. The concern with this is that the supply of 'suitable' sites is limited, especially in the vicinity of large cities. The supply of possible landfill sites is clearly not an issue. The key word in this concern is 'suitable'. Often the scarcity issue is more in terms of the resources used to transport the waste from its origin to a landfill site. And there is also a scarcity of local communities that are willing to accept the location of a landfill. This is especially true when affected communities are provided with no compensation for their loss of environmental amenity.

For both of these scarcities – transport resources and community acceptance – costs are involved and signals (regarding relative scarcity as reflected by the costs), need to be sent to those who are generating waste. This adds to the difficulty of trading off between waste creation and illegal disposal.

It is worth reiterating that the creation of waste is not the intention of people in their consumption or production choices. Waste is something people try to avoid. Better to have resources that are valuable. Better not to have the costs of disposing of the waste. The concerns raised in the context of landfill should therefore not be focused back on to the creation of waste but rather the disposal thereof. For instance, the banning of some forms of waste (for example, plastic bags) is unlikely to be the best solution to achieving a reduction in the number of plastic bags that litter the streets. The banning of plastic bags means that the service being provided by those bags is either lost or is having to be carried out by an alternative higher cost option. So in Delhi, where plastic shopping bags have been banned, people have to find alternative solutions to their storage and carrying problems. They will be more expensive and/or less effective because they were not the chosen option prior to the ban. That implies more and scarcer resources being used. The rag pickers will lose a source of income from their 'recycling' activities.

The experience in Ireland when plastic shopping bags were subjected to a 15 euro cent tax was that the sale of plastic bag bin liners increased dramatically.[23] Apparently, many households were 'recycling' their shopping bags around the home. With that supply cut-off, an alternative supply in the form of new plastic bags had to be found. Similarly, paper bag consumption increased, with implications for the use of other environmentally notable scarce resources, including energy and forests.

The answer to the problem created by plastic bag waste – the unsightly scattering of them through the landscape – is in the management of the waste stream rather than in their production and consumption.

A 'little green lie'?

The initial impressions most people have of waste-reduction campaigns and calls for greater recycling are that they must be good. After all, waste is a bad thing and the less we have of it the better. Those advocating the 'reduce, reuse, recycle' mantra likewise are motivated by what seems to be the best interests of society. They are popular causes. The problem again is that 'all that glitters is not gold'. A closer analysis of the waste issue indicates that there are likely to be unintended negative consequences.

Waste is not produced for its own sake. It is an unwanted by-product of production and consumption activities that make people better off. If policies are introduced to restrict waste, then necessarily those policies will restrict the production or consumption of something that people like. That will make them worse off.

Policies that encourage waste reuse or recycling are also likely to make people worse off because they require the use of scarce resources and hence impose costs. The goal of reducing scarce resource use may not even be achieved. The outcome will merely be a re-allocation of demand for resources away from the one that is targeted by the waste restricting 'rules'.

---

23 http://www.guardian.co.uk/environment/2009/aug/11/plastic-bags-welsh-assembly

If resources are 'saved' by waste policies such as recycling requirements or subsidies, those savings are likely to come in the distant future. Only after many years of using the cheaper virgin resources will the recycled resources come into play. With such long time periods involved, there are significant uncertainties about the development of possible substitutes and hence whether the 'saved' resources will be used at all. The value of those future gains is thus likely to be small now.

However, society faces a real quandary in terms of how it manages solid wastes. To make those who create waste recognise the full costs of their actions, prices for the use of landfills need to reflect the full costs of operation, including payments to compensate local residents for loss of environmental amenity. However, such a pricing strategy is likely to mean more waste will be disposed of through littering and illegal dumping. Many authorities therefore choose to implicitly subsidise the collection and disposal of household waste as a lower cost means of reducing the negative environmental consequences of littering and illegal dumping. To assist this, community programmes of education regarding littering mean that there is more community resistance to waste disposal outside of landfills. The actions of waste reduction community groups are often supportive of these educational efforts. The danger to society comes when their calls shift from a focus on improved waste disposal to one of reducing waste production. Then the prospects of higher costs and less efficient use of resources become real.

7: Using Resources 'Efficiently'

Proposition: Resources such as water and energy should be used 'efficiently', whatever it costs.

Resources such as water and energy are scarce. Use of these resources needs to be minimised so that future generations will have enough. Governments should invest in technologies that will ensure the least amount of energy and water is used in producing goods and services.

BUT

Investing in 'efficiency' measures often means using other scarce resources as substitutes for energy and water. A 'false economy' results because the other resources, including labour and capital, may well be scarcer than energy and water.

Be efficient!

It seems common sense. If there isn't much left of a resource and that resource is important in the production of things we really want, then we should make sure we get maximum productivity out of the remaining resource. Otherwise, we'd be 'wasting' what was left of the resource.

Following this logic, governments faced with resource scarcity 'crises', such as for water when there is a long running drought or for energy when oil prices 'spike', often seek ways to improve the 'efficiency' with which we use the resource. By increasing resource-use 'efficiency', the idea is to save the available resource stocks so that more will be available for future use.

For instance, in Australia's recent ten-year-long drought, governments provided subsidies for people to install shower heads that limit the amount of water that can flow through and clothes washing machines designed to use as little water as possible. Some schemes involved 'new for old' shower heads.[1] Farmers were provided with assistance to install water-saving technologies (including computer controlled drip irrigators)[2] and water supply authorities received funding to line open irrigation channels (which are subject to water loss from seepage and evaporation).[3]

In response to concerns about energy shortages (as well as greenhouse gas emissions arising from energy production) numerous jurisdictions around the world including the EU,[4] California,[5] and Australia[6] have either outlawed the use of conventional, incandescent light bulbs or legislated standards so that 'low energy' compact fluorescent bulbs have become widely used. Rebates are paid for the installation of energy-saving solar hot water systems.[7] Electrical appliances such as refrigerators and air conditioners, by law, are given 'star ratings' to advise consumers of their energy consumption levels.[8] 'Energy audits' provide free advice to households and businesses.[9] In some jurisdictions, such as the Australian Capital Territory, houses cannot be sold unless their owners have had them assessed by an independent auditor for their energy efficiency and given a 'star rating'.[10]

---

1   http://www.livinggreener.gov.au/rebates-assistance/vic/showerhead-exchange
2   http://www.environment.gov.au/water/programs/srwui/on-farm.html
3   http://www.victoriasfoodbowl.com.au/irrigation/overview.cfm
4   http://www.nytimes.com/2009/09/01/business/energy-environment/01iht-bulb.html
5   http://www.brighterenergy.org/21541/news/heat-efficiency/inefficient-100w-light-bulbs-banned-in-california/
6   http://www.climatechange.gov.au/what-you-need-to-know/lighting.aspx
7   http://www.environment.nsw.gov.au/rebates/ccfhws.htm
8   http://www.energyrating.gov.au/
9   http://www.environment.nsw.gov.au/sustainbus/energyauditing.htm
10 http://www.actpla.act.gov.au/topics/property_purchases/sales/energy_efficiency

The aim of all of these policy measures is to make the available scarce resource 'go further'. They are all phrased as 'efficiency improving' measures. On the surface, they would seem to be sensible measures. After all, efficiency is something most of us strive for. We are praised for being 'efficient' at our jobs. Companies impress their shareholders with claims of improved 'efficiency'. And, in contrast, 'inefficiencies' are frowned upon. But what exactly is meant by the term 'efficiency'?

When is being efficient not efficient?
There are two ways of looking at the concept of efficiency. Hence there is a potential for confusion. The first interpretation is known as 'technical efficiency'. It's what engineers, production managers and most biophysical scientists refer to as 'efficiency' and it relates to the amount of output produced for the amount of input applied. If more output is achieved for a given input, greater technical efficiency is said to have been achieved. Similarly, if the same output can be achieved using less of an input, it is more efficient.

The problem with this interpretation is that there are many ways in which technical efficiency improvements can be achieved. For instance, more output can be achieved with the same input of a particular resource if more of another resource is used. This is because resources are able to substituted for each other either in the production of goods and services or in the process of consumption. Farmers may be able to increase the efficiency of water use in the production of an irrigated crop by increasing the amount of high-tech irrigation equipment they use. By applying water to their crop using computer-controlled drip feeds or centre-pivot irrigators, they can use less water to achieve the same output as they previously produced using simple (low-capital input) flood irrigation techniques. Similarly, electricity consumers can lower their lighting energy use, but it might take a change in the equipment they use from conventional incandescent bulbs to compact fluorescent globes. In effect, the investment in different globes and different irrigation systems involves a substitution of capital equipment for energy and water.

With technological advances, there is the prospect that more of these technical efficiency improvements will become available. For instance, the genetic engineering of plants to require less water may see outputs maintained with far less irrigation water inputs. Washing machines have been developed that use nylon beads to remove dirt from clothes instead of water.[11] New light bulb technologies such as LED (light emitting diode) and quantum dots are already superseding compact fluorescents in terms of their technical efficiency.[12]

The process of input substitution leading to improved technical efficiency means that the use of one scarce resource (for example, water/energy) is replaced by the use of another set of scarce resources. The question remains as to whether or not this substitution should occur. This is the realm of the second interpretation of the efficiency concept: 'economic efficiency'. This revolves around the choice between the different combinations of resources that can be used to create the desired consumption or production outcome.

The bottom line to economic efficiency is that you choose the combination that makes people best off. When the outcome of differing input combinations is the same, all that is needed to satisfy the economic efficiency criteria is that the inputs used are those which cost the least. In other words, one combination of inputs is more efficient than another combination of inputs if it has a lower cost.

This logic is consistent with the role that prices play in signalling to users the relative scarcity of resources.[13] The combination of resource inputs that is the cheapest will also be the combination of resources

---

11 http://www.economist.com/node/13892738?story_id=13892738
12 http://www.economist.com/node/17030544
13 This requires that prices are formed in competitive markets for resources that have well defined and defended property rights. More on the context where this may not be the case is considered later in this chapter.

that is the least scarce. Economic efficiency therefore implies using resources that are (relatively) the most abundant.

What this means is that an option which is technically efficient may not be economically efficient. If the resources that are used in place of the water/energy resource that is the focus of a technical efficiency drive mean that the overall costs go up, then technical efficiency is being achieved but the outcome is not economically efficient. So with greater technical efficiency, less of the water/energy resource will be used and that will mean some cost savings. However, if the replacement resources (capital equipment, for example) cost more than those savings, the change will be economically inefficient, as people will be overall worse off.

Policy implications

People acting in their own best interests are unlikely to choose options that are not economically efficient. If the same outcome can be achieved at a lower cost, the lower cost option will be chosen. Hence, a farmer will not bother investing in a drip irrigation system unless it costs less than what can be saved in water bills. Compact fluorescent bulbs will only be chosen if the extra cost of that type of bulb can be paid for by the electricity bill savings that result.[14] However, government policy designed to encourage technical efficiency may cause people to choose economically inefficient options.

Of course, they won't choose those technically more efficient yet economically less efficient outcomes if it makes them worse off. They will only do so if governments induce them to make the change through various policy mechanisms. Subsidising the cost of the substitute resource is a common option used to do this. So governments offer subsidies for the installation of low water use/energy use appliances and meet the costs of new irrigation systems for farmers. These payments from the

---

14 This also assumes that the quality of the light from the alternative systems is equivalent.

public purse give the incentives for people to use water/energy resources in more technically efficient ways. However, just because those changes make them better off, it does not mean that the changes are economically efficient. Taken from a society-wide perspective, people are made worse off by the changes because the changes are, overall, more costly.

Similarly, policies that involve regulations limiting the use of less technically efficient resource combinations (such as bans on incandescent light bulbs) induce people to make changes that they otherwise would rather not make. The higher costs they have to pay in order to move from the restricted option to ones that the government is seeking to promote in order to 'save' water/energy are a measure of how much worse off they are. The regulations are therefore economically inefficient.

Inefficient for a reason?
What could justify a government introducing policies that apparently make people worse off?

One possibility is that the price of the resource that is being targeted by the government's 'efficiency drive' (for example, water, energy) is somehow not reflecting its true scarcity value. If the resource price is somehow 'too low', then the signal it gives is that the resource is more abundant than it really is. Hence, people will be less inclined to choose technically more efficient options that are also economically more efficient when true scarcity prices are taken into account. Governments may therefore seek to subsidise the more technically efficient option in an attempt to rectify the price-signalling problem. For instance, if property rights to water are not well defined and if competitive markets have not formed to allow the generation of accurate price signals, then water prices may not reflect the true scarcity of the resource. Governments may seek to redress the resultant over-use of the resource through encouraging water efficiency measures. Under the same logic, government may seek to redress the 'too low' price of a resource by imposing a tax on its use. The change in price that results would act to encourage the use of more 'efficient' resource-use options.[15]

Another possibility is that people may be poorly informed about the availability of technically more efficient options or may not be aware of the cost savings they provide. Adoption of energy-saving lighting may, under this proposition, be limited because people are not aware of the savings over time that could be enjoyed. People may decide to be ill-informed because collecting and processing the information necessary to make a choice is expensive relative to the extent of any benefit they expect to enjoy as a result of being better informed.

It may also be true in this case that people don't care so much about the energy cost savings because they will only be enjoyed in the future. The rate at which people 'discount' those future savings may be 'too high' when considered from a society-wide perspective. This would imply that people acting on their own are 'short sighted' compared with the way society as a whole would choose. This is because, as individuals, people take into account factors that make future savings risky, but these factors do not apply to the broader society. For instance, as individuals, we may not live long enough to enjoy the savings, but for society as a whole, the death of individuals doesn't make any difference to society's continued existence.

The cost savings from using less of the targeted resource are usually generated by individuals making investments in capital equipment. This means that cost savings spread out through time into the future are weighed up against the immediate capital cost. The extent of the capital cost may also be seen as a barrier to the up-take of technically more efficient options. If that up-front investment is high relative to the income available to pay for it, people may find it difficult to arrange a

---

15 The choice between a tax and a subsidy has implications for the distribution of wealth in society. While a tax makes the resource user worse off, a subsidy improves their lot at the expense of the general taxpayer. Politically, the subsidy option is more attractive to governments especially when the resource users are predominant in marginal electorates.

loan. The transaction costs of setting up a loan may be too high, or credit may simply be unavailable, because of assessed ability to repay a loan.

Another rationale for regulating, taxing or subsidising to encourage 'efficient' use of a resource (water/energy) is to deal with the environmental impacts of their use. For instance, banning incandescent globes has been justified because of their higher use of energy and hence their generation of more greenhouse gas emissions. The argument is that with no market for these emissions, no price is paid for them and thus those environmental costs are not factored into the choices people make between alternative lighting modes.

How well these rationales stand up to scrutiny is considered in the next section.

Signalling scarcity

The first rationale for government intervention concerns poor price signals. The approach of governments' subsidising or regulating to achieve 'better choices' than those made under distorted prices is problematic for a number of reasons. First, it requires government to work out the 'right' level of subsidy or tax or the 'right' degree of regulation in order to generate those 'better choices'. This is a difficult task given the informational requirements and the prospect of fast-moving variations in the conditions that determine what is 'right'. There is also the need for government to fund subsidy payments. This in turn requires the collection of taxes and the host of costs that are involved in the administrative process, let alone the disincentives which taxes have on individual work effort.

Whenever subsidies are paid to make one option relatively more attractive, it is always worthwhile remembering what impact the subsidy has on the resource-scarcity signalling function of prices. The subsidy payments effectively lower the price that is signalling the relative scarcity of the resources that are acting as substitutes for the resource being targeted by the 'efficiency drive'. With the lower price, more of those substitute resources will be used. They, too, are scarce and the altered scarcity signal forms a distortion to the allocation of resources that can end up making society worse off.

Paying subsidies, levying taxes and regulating use are also 'indirect' approaches to addressing the issue that is the cause of the problem: Poor price signals. The more straightforward solution to that problem is to address the reasons why competitive markets for the targeted resource have not formed.

An example in Australia has been the development of well-defined property rights for irrigation water. This has enabled the trading of irrigation water in competitive markets with adjunct assignments of water to be used to generate environmental benefits. The consequential increases in water prices have meant that more and more farmers have decided to invest in water-saving technologies without subsidy or regulation. A tax on water use was not required because the market ensured that relative resource-scarcity signals were delivered to farmers. As a result, irrigators have realised that the cost savings associated with reduced water use are now sufficient to justify investments in technically more efficient irrigation methods.

It should be noted, however, that the resource-scarcity signals provided by the market are in part due to the government's determination of the amounts of water to be held back from irrigation so that they can be used to provide environmental benefits. Because of the costs of establishing property rights over those environmental benefits, they are yet to be included directly in the market mix without government involvement.

Information

The second rationale for government involvement relates to poor information. This is a 'paternalistic' approach because it implies that government knows better than individuals. The difficulty with this approach, however, is that the circumstances of each individual vary and government is in no position to be able to say what is 'best' for all its constituents. For instance, banning incandescent light bulbs on the basis of people not fully understanding the cost savings they provide ignores people's differences in terms of the amount they want to pay for lighting and the extent to which they want to trade off future cost savings against current expenditure on new, more expensive bulbs.

On top of this there are likely to be differences in the ways people perceive the light outputs of different types of bulbs. People who have high values for the banned incandescent light bulbs (possibly for the characteristics of the light they create) and who hate the compact fluorescent alternative (perhaps because of their 'flicker') will bear larger costs than those who are indifferent between the two light sources. With such variations in tastes, the 'one-size-fits-all' strategy of banning conventional bulbs does not allow for different people to respond to relative resource scarcities in ways that best suit their situations and circumstances.

A better strategy for dealing with concerns about ill-informed choices is for government to require information to be supplied with products. One example of this type of policy is the energy rating schemes under which appliance manufacturers are required to give information about the (energy and water) running costs of their products; another is motor vehicle producers being required to state the fuel consumption of their models. The rationale here is that this information is known to the manufacturers but is costly for potential buyers to collect and process. Ensuring that manufacturers share their product knowledge with consumers is a low-cost way of generating buyer awareness. It allows consumers to make well-informed choices that best suit their individual circumstances, likes and dislikes.

This includes the choice to trade off future running costs against current purchase costs. Once people know the current and future cost consequences of the available options they are in the best position to make a choice. Forcing the choice – for example, through the elimination of the incandescent bulb option – can cause some seemingly undesirable outcomes. For instance, well-off people may find it easier to fund an investment in higher cost energy-saving globes compared with poorer people who have higher priorities for current food and clothing purchases. Thus a ban on the cheaper-to-buy incandescent bulb may be particularly harmful to poorer people.

Credit

The third rationale for government intervention involves a further barrier to people investing in efficiency-enhancing options. It is the availability of credit. Overcoming the obstacle of an initial capital cost may be very difficult for people with low (or negative) disposable incomes. What do these difficulties imply?

First, there will always be an income constraint. Second, people may have preferences that favour the current consumption of goods over an investment that saves future energy or water costs. Hence they will use their available income for current consumption. To extend their consumption beyond what their current incomes will allow, people would need to borrow. But borrowing is necessarily costly: Those who lend the money will want to be paid interest for giving up their use of their funds for the period of the loan. Depending on the chances of the borrower paying back the loan, these interest payments will vary: The higher the risk of default, the higher the interest rate. In addition, there are the costs associated with the administration of the loan. Banks and other lenders use scarce resources to facilitate the transaction between borrower and lender. These costs must also be paid by the borrower.

Because of these costs, for some people, the apparent benefits associated with lower future energy/water use costs due to the investment in 'efficiency' may not be sufficient to justify a loan. This holds with greater force as interest rates rise. Hence, those people with more chance of defaulting on a loan (low future income, high future expenses) will be less likely to find a loan worthwhile, or less able to find a lender willing to take the risk.

Given this insight, the banning of incandescent light bulbs is likely to do more harm to those who have low disposable incomes and with poor credit ratings. In other words, the policy may end up being regressive: it costs the most for those who are least able to afford it.

The environment
The other rationale for subsidising/taxing/regulating for efficiency is the environmental impacts of the 'inefficient' alternative. Bans and less strident regulations that limit use are likely to be 'blunt instruments' in trying to achieve environmental improvement goals. The particular way of reducing emissions, for example, chosen through the enforcement of a ban or regulation may not be the cheapest way to achieve the environmental protection goal. For instance, banning incandescent light bulbs may reduce energy consumption and hence greenhouse gas emissions, but it may not be the most cost-effective means of doing so. This is especially so given the differences in costs that a ban implies across different people. For some people it may be cheaper to sequester carbon emissions by planting trees or even to pay others to do so. The imposition of a ban does not allow flexibility and adaptability to arise in dealing with an environmental issue.[16]

The banning of plastic bags is another case. While the justification for the ban has largely been made in terms of waste reduction, there is also an element of 'efficiency' in the argument: If disposable plastic bags can be replaced by re-usable (but more expensive) cloth bags, the available resources are being used more 'efficiently': Nothing is being wasted if the bag can be used over and over again. The banning of plastic bags is a heavy handed way of dealing with the waste issue that precludes the development of innovative ways of dealing with the waste so that it

---

16 In contrast, subsidies and taxes do provide the resource user with the flexibility to decide how they will deal with the changes in relative prices that result. Bans and use regulations offer no such flexibility and are therefore likely to be more costly to resource users.

becomes a valued resource. It ignores the diversity of preferences and incomes that exist across the community.

Another possibility associated with 'efficiency'-promoting policies is that, by seeking to improve the environment in one dimension, they can have the potential to create other environmental problems. For example, the disposal of compact fluorescent globes presents difficulties because they contain the heavy metal mercury. In some states of Australia, the disposal of mercury-containing light bulbs to landfill has also been banned because of the toxicity of mercury.

The same problem applies to hybrid and electric cars. These cars use less petrol to cover a given distance than comparably dimensioned petrol-only cars. Hence they can be classified as being more 'technically efficient' in terms of petrol consumption. They are, however, in general, more expensive to manufacture and thus for consumers to buy than comparable conventional cars. Hence their overall economic efficiency can be debated: Are the additional scarce resources used to make a hybrid car worth more or less than the petrol that is saved? For many consumers it would seem that the answer is 'no' because the Australian Government is subsidising the development of the hybrid Toyota Camry to the extent of $35 million[17] in an attempt to encourage more people to take on the more technically efficient hybrid technology. While hybrids do have lower greenhouse gas emissions profiles than conventionally powered cars, they give rise to the additional environmental problems associated with the disposal/recycling of their nickel-metal hydride battery packs.

Environmentally motivated efficiency policies may also prove ineffective. An example is the move to line irrigation channels to prevent seepage to groundwater. The logic of the efficiency argument goes as follows. With less water 'lost' from the channels, less water needs to be diverted from a river to satisfy irrigators' demands. This 'saved' water

---

17 http://www.caradvice.com.au/15946/toyota-recycling-hybrid-batteries/

can then be reallocated to the environment: For example, for watering wetlands and maintaining the health of fish populations. The problem with this logic is that the water is not 'saved' but is rather reallocated around the river basin. With the lining in place, less water percolates through to the water table. With lower inflows, the characteristics of the groundwater table will change. So while more water may be allocated to downstream wetlands, ecological systems that are dependent on the existing groundwater table will be adversely affected.

Efficiency encouragement can also have 'rebound' effects. With greater efficiency in the use of a resource, its value as an input is enhanced. Because the resource is more productive, people choose to use more of it. Ironically then, the efficiency improvement does not lead to less of the resource being used. To the contrary, more of the resource is used.

A 'little green lie'?

Efficiency would appear to be akin to 'motherhood and apple pie' — something that just has to be good. It would seem, however, that when the initial appeal of the concept is put to one side, there are some elements of efficiency that are not necessarily attractive. The problems arise primarily because efficiency can be viewed from a one-dimensional basis. And that dimension is focused on a specific resource that has been identified as warranting special consideration. So we see water efficiency and energy efficiency as being the focal points of policies.

Many of these efficiency-enhancing policies involve subsidies or regulations, including outright bans on some products that are deemed to be too 'inefficient'. Subsidies are frequently the preferred mechanism that governments use to encourage 'efficiency' because they allow politicians to confer 'favours' on interest groups and don't run the risk of alienating voters that is inherent in 'inefficiency' taxes.

The encouragement of 'efficiency' comes from a variety of sources.

First there are the environmental groups who focus on the particular resource they consider to be 'at risk'. Their strong preferences for the resource of interest tend to lead them away from considering the impacts of 'efficiency' policies on other resources. They see these resources as relatively less valuable and so are personally willing to make the trade-off between the substitute resources and their resource of interest. This is particularly the case if they personally do not have to pay the costs of the substitute. So if more water can be made available for environmental flows in a river system through governments paying the costs of efficiency improvement schemes, those with strong environmental preferences will be supportive. That way they can have more of the environmental benefits they enjoy with the costs of providing them being spread across the taxpaying population base.

The key point to this explanation is that those of an environmental persuasion seeking the promotion of efficiency enhancement policies are likely to have preferences for the resource in question that are stronger than the rest of the population. Their perceptions, therefore, of relative resource scarcity are likely to be different from the average perception across an economy. They would thus see the price signals being generated from people's actions across the whole population as fundamentally undervaluing their resource of concern. They seek opportunities to have government policies installed that reflect a stronger preference for the target scarce resource. This is despite the higher individual, personal costs which they cause. Pushing the 'little green lie' helps in that regard, as it takes the focus away from the costs of using more of the substitute resources.

It is also apparent that others will gain from the propagation of the efficiency 'little green lie'. These 'winners' from efficiency-seeking policies are those who will profit from the subsidies and regulations. No doubt the manufacturers of compact fluorescent light bulbs are significantly better off as a result of the banning of incandescent bulbs. No doubt farmers who have the investment costs of high-tech irrigation equipment paid for by government are better off given the reductions they could expect in their future water bills. Even households who have their showerheads replaced by water-saving substitutes at public expense will have lower water bills without the up-front equipment investment costs. The subsidies received by Toyota to develop its hybrid car range were no doubt welcomed by the company's shareholders. And of course, those whose careers are enhanced by the successful introduction of their technological breakthrough (the scientists and engineers) are also likely to be key supporters of 'efficiency' enhancement.

The subsidy payments and regulations all provide an opportunity for interest groups (consumers and producers) to achieve some specific improvements in their well-being at the expense of the wider community. The costs of the policies are spread thinly across the populace, or at least taxpayers, and the benefits are concentrated within relatively small groups of people or businesses. This is a formula for 'little green lies' to be very successful in convincing politicians to introduce (technical) efficiency-enhancing policies. Votes will be won (and lost) by providing (or not) benefits to concentrations of people. The policy decision has such an impact on these people that their voting behaviour is affected. However the extra cost burden caused by the policy, because it is spread thinly across all taxpayers, is likely to go unnoticed. In essence, the 'voting power' in 'political markets' is usually distributed differently from 'dollar votes' in actual markets. Where markets can be formed, dollar votes are much better measures of individuals' true valuations of alternatives. The challenge is to develop innovative institutional structures that lower the costs of market formation.

8: The Infinitely Valuable Environment

Proposition: The environment is of infinite value and must not be harmed.

The environment provides us with our 'life-support-system'. Without it we cannot survive and so we should protect it at all costs.

BUT

Without the environment we could not exist and so its absolute value is infinite. However, that is not the relevant question for policy. Changes to the state of the environment yield finite benefits and costs that need to be traded off when making policy decisions.

Infinity and beyond

There is no doubt that the environment is valuable. If we take a very broad definition of the environment to encompass the world in which we live, it becomes abundantly clear why it is so valuable. It provides us with food, water and all the natural resources we rely on for production and consumption. It regulates the climate we live in. It absorbs our wastes and gives us opportunities for recreation, relaxation and contemplation. Without the environment, humanity could not survive.

This logic provides support for the proposition advanced in this chapter. And the logic is very difficult to dispute. Humanity is an integral part of the wider environment and although the environment is affected by humanity it is not dependent on our continued existence: Without people, the world would be a very different place but it would still exist. In contrast, not only does the environment impact on humanity, but there is also a relationship of absolute dependency between humanity and the environment.

If we define 'value' as the amount of money (in its role as a measure of people's ability to access resources) that a person is willing to give up in order to prevent the environment being destroyed, then, with survival on the line, the value of the environment is likely to be all the money that people have. What would be the point of being able to access resources if there was no environment to keep us alive? This value is obviously a very large number if not, technically, an infinite amount.

Given that the environment is made up of resources that people can access, the concept of people being willing to give up access to the resources they need to survive seems illogically circuitous. The end point of the logic is people giving up all access to the environment so as to protect it. Because access to the environment ensures humanity's survival, the ultimate 'sacrifice' for the environment would be to give up on that continued survival. In human terms that would mean an assignment of value to the environment as close to infinity as anyone would care to define.

What's the question?

Although the value of the environment from this perspective is interesting conceptually, it is not practical or useful in terms of the actual decisions we must confront on a daily basis. Even the most strident environmental advocate is unlikely to be willing to give up his or her own life or the lives of others to improve the environment or to protect it from damage, let alone advocate the complete extermination of humanity.[1]

---

1   Although advocates of reducing the world's population to 'sustainable' levels are often unclear as to how such reductions are to be achieved, at least in the short term. The Voluntary Human Extinction Movement (VHEMT) advocates that 'phasing out the human race by voluntarily ceasing to breed will allow Earth's biosphere to return to good health': http://www.vhemt.org/

However, the irrelevance of the 'infinite value' concept is useful in pointing towards a more useful way of looking at the question of the environment's value.

In terms of practical environmental decision-making, no-one is being asked to sacrifice their lives to save the environment from destruction.[2] It's a choice that isn't contemplated because the choices made by people do not involve the complete destruction of the environment. Rather, the day-to-day decisions that people make may cause some reduction in the goods and services provided by the environment. Wholesale destruction is not part of the choice equation.

Instead, the relevant question regarding the value of the environment is what are people willing to give up in order to have a little less environmental harm or a little more environmental good?

The reality is that, every day, people make choices either to harm the environment or to improve it. Our choices to drive cars, turn on the lights, write on paper, flush the toilet and almost everything else we do all involve negative environmental impacts. But we also decide to do things such as pay premiums for phosphate-free detergents and dolphin-safe tuna, volunteer our time to plant trees and donate to 'save the whale/forest/wetland/...' campaigns. These are options we choose that have positive environmental impacts.

The key feature of all these choices, both positive and negative for the environment, is that they involve small changes to the condition of parts of the environment; these are changes 'at the margin'. It is the value of these environmental impacts that is relevant to decision-making and not the value of the environment in total.

---

2   This does not deny that some people voluntarily place themselves in great danger – lying in front of bulldozers, sailing in front of whaling vessels, etc. – in support of environmental causes.

How much?

We are not faced with choices that involve the fate of the environment as a whole. Rather, we have to decide between a host of options that each have small impacts on the condition of the environment. Hence we need a way of working out just how much value we place on small changes to the environment. We need to know the extent of the costs of the environmental damage we cause when we decide to drive the car or turn on the lights so that we can decide if the benefits we enjoy from making those choices are worth those, and any other, costs we have to bear. Likewise, we need to be able to weigh up the strength of our preferences for the environmental gains we enjoy from harming fewer tuna or setting up more wetland protected areas against the costs we bear from doing so.

When we make choices, we do so on the basis of our understanding of the balance of costs and benefits offered by each option. That means information on the costs and benefits of our choice options is routinely collected and processed in the course of everyone's day-to-day activities. Choices to buy or sell in markets are advantaged because of the information that markets provide. The price of a good or service in a competitive market signals the value of the embodied resources in their next most valuable use. Bids from competing users of the resources make sure that's the case. In deciding to make the purchase, we have to decide if the value we will gain by having the good or service is greater than the price and therefore greater than the value held by other potential users. If the resources involved become scarcer, only those users who have relatively high values will remain as buyers. Prices will rise to reflect the changing situation. Decisions taken will then reflect the changed prices.

Market prices thus perform the function of signalling the values held for goods and services by buyers and sellers. Specifically, they indicate these values 'at the margin': They show how people value small changes in the supply and demand of goods and services.

Markets therefore provide a mechanism whereby individuals are able to make the trade-offs between costs and benefits of options. They facilitate the coordination of people seeking to make themselves better off through their consumption and production activities. Resources move from relatively low-value uses to higher value uses through market trade. The social coordination so provided is made possible through prices signalling information about the relative values the people place on the host of resources that area available.

But do markets provide information on the value of changes to the environment caused by our choices? Do markets signal the relative scarcity of environmental assets?

The answer is: 'not always'.

Where property rights to environmental goods and services (ownership) cannot be defined, or are very costly to define, or where any defined right cannot be defended or is very costly to defend, markets are unlikely to form. In such circumstances, the costs of market establishment are too high relative to the gains from trade that could be expected to arise through market exchange.

And herein lies a potential problem. If markets are not forming to deliver the correct signals regarding the scarcity of environmental assets, the inference is that the value of the environment is zero.

So while we know that the value of 'marginal' changes to environmental assets is not infinity, we also know that it is not zero. Because there are many potential uses for environmental assets and because there are in most circumstances insufficient of those assets to satisfy all those uses, the assets are scarce and hence valuable.

The consequences of changes to the environment being infinitely valuable are dire. We would never be able to use the goods and services that the environment offers because no use would ever be able to generate a benefit sufficiently large to outweigh the cost of losing some of the environment. But so too are the consequence of a situation in which changes to the environment are of no value. The environment would be run down through over-use and we would lose many of the (valuable) goods and services it provides.

Non-price signals
Neither the zero nor the infinite value signal for changes to environmental assets is correct and both will lead to resource-use outcomes that are not in the best interests of society. Both should be avoided.

To make sure people do receive appropriate signals about the environmental costs and benefits of their actions there are fundamentally two ways forward. Both involve action being taken at a communal or centralised level, often with governments taking the lead, rather than at the individual or decentralised level. The first is to make sure that markets do form for environmental assets and the goods and services they provide. Here, collective action may be able to generate economies of scale sufficient to lower the transaction costs of establishing and enforcing property rights so that they are less than the gains from trade that result. With markets established, prices for accessing environmental assets will form and scarcity signals will thus be transmitted.

Where the transaction costs of defining and defending property rights to allow market formation remain prohibitively high, other collective action may be warranted. However, it may also be the case that the costs of collective actions such as governments stepping in to protect environmental assets (for example, legislation to set aside national parks, regulations to limit pollution, subsidies paid for the private provision of environmental services) are greater than the benefits so generated. If this is the case, the government action would make society worse off. To avoid that situation, government environmental policies need to be assessed in terms of their benefits and costs. This is known as regulatory impact analysis or assessment.[3] In some juris-

---

3   For the UK: http://www.parliament.uk/site-information/glossary/regulatory-impact-assessment/ EU: http://ec.europa.eu/governance/better_regulation/impact_en.htm Australia: http://www.finance.gov.au/obpr/ris/gov-ris.html

dictions this is a legislative requirement, with the OECD providing guidelines of 'best-practice'.[4]

Values without prices

The assessment of potential collective/government interventions requires the estimation of the benefits arising and the costs borne. In order for the two to be compared – and hence to enable the formation of a judgement as to the net impact of an intervention – the estimations need to be in the same unit. For environmental costs and benefits where markets have not formed, this means estimating values without the evidence provided by market prices.

An increasingly prevalent way of doing this is through what are known as 'stated preference' surveys. In these surveys, people are asked questions designed to gain an understanding of the strength of their preferences for specific environmental changes. This usually involves survey respondents making trade-offs between their having more of an environmental good (or less environmental damage) and money. Value is thus defined in terms of respondents' 'willingness to pay' for environmental changes.

This is a controversial approach for at least two reasons. These are set out in the next two sections.

It shouldn't be done

First, those who take the stand that the environment is of infinite value suggest that it is morally inappropriate even to attempt to put a monetary value on damage being done to the environment. At the extreme end of this argument, a parallel can be drawn between the question of the willingness to pay to avoid environmental harm and the willingness to pay to stop mass murder.

This unwillingness to consider trade-offs can also be driven by a reluctance to associate money and the environment: Anyone

---
4   http://www.oecd.org/dataoecd/21/59/35258828.pdf

considering a trade-off between environmental condition and money would be lumped into the same category as someone who would be 'willing to sell their own grandmother'.

There is also a 'rights-based' interpretation of the 'morally inappropriate' argument: Payments to avoid environmental damage should not be required because people have a right to the situation where there is no environmental harm.[5] Just as payment to stop murder should not be required because people have the right to live without fear of such atrocities, so it is argued that people should not be 'held hostage' over the threat of environmental harm.

The strength of these arguments is diminished as the scale of the environmental harm being contemplated is reduced. The analogies with mass murder and grandmother-selling lose some relevance when the trade-offs involve the felling of an additional hectare of trees in a region where 1,000 hectares of forest remain. Put simply, in most cases, stated preference questionnaires don't go anywhere near the extremes implied by the sale of relatives and crimes against humanity. So while broad ethical frameworks may be useful in decision-making about issues such as murder and slavery, they do not provide much useful guidance in most practical environmental management issues.

That said, it is also apparent that the extent of the potential environmental harm caused by some options for using resources is not trivial. And the cumulative effects of many small decisions can eventually breach some threshold so that overall they become substantial.[6] In such cases, and especially when the environmental harm is 'irreversible', as is the case with species extinction, then the simple

---

5   The implication of this vestment of rights is that compensation payments should be made to those who must endure environmental harm.

6   The 'rivet-popper' analogy comes into play. The removal of one rivet from the wing of an aeroplane may make it lighter and fly faster and more economically, but eventually the loss of an extra rivet will ensure that the plane will crash. See Ehrlich, P. and A. (1981), *Extinction: The Causes and Consequences of the Disappearance of Species*. Random House, New York.

trade-off notion becomes more complex. Does that mean that when more substantive environmental harm is being contemplated or when the small amounts of environmental harm accumulate to reach a level that breaches some form of threshold that the environment does attain 'trump status' and trade-offs are suspended?

The problem with this logic is that there are always two sides to any trade-off. Yes, the environment may be harmed by a choice, but what happens if the environmentally harmful option isn't taken? Perhaps the alternative would involve greater poverty for people with associated increased incidences of illness and infant mortality. Then the question of which moral imperative – preventing environmental harm or preventing human suffering – is the greater has to be raised and a trade-off once again emerges.

This notion of suspending the trade-off way of thinking when a threshold of environmental damage is reached was first contemplated by Ciriacy-Wantrup.[7] He established the notion of a 'safe-minimum standard' in the condition of the environment. An example would be a threshold number of individuals of a species under threat. Once the safe minimum standard was reached, further declines in the condition of the environment were only to be chosen if the costs of not doing so were 'intolerably high'. Defining exactly what is meant by 'intolerable' necessitates the contemplation of what people would be willing to pay to avoid the environmental harm as compared to their willingness to pay to go without the benefits created through the use of the environmental assets.

What is apparent at this point is that trade-offs are very hard to avoid.

Taking money out of the trade-off equation is also difficult. Even though some people may regard the environment to be on a different plane to money, there is plenty of evidence to show that the two are frequently intertwined. Markets have formed for some environmental

---

7   Ciriacy-Wantrup, S.V. (1952), *Resource Conservation: Economics and Policies*. University of California Press, Berkeley.

goods, or at least some markets involve elements of the environment. For example, it is possible to buy a better living environment by purchasing a more expensive house that is located away from noise pollution sources and close to pleasant parks. An analysis of house prices can even be used to infer the amount people are willing to pay for less noise pollution and better access to parks. Another example is to be found in the donations made by people to support environmental conservation causes, including the purchase of land for the establishment of nature protection areas.[8]

The rights-based objection to making willingness to pay trade-offs in stated preference surveys has merit. Where there are well established, if only implicit, rights to the provision of a level of environmental good, then asking survey respondents to pay to avoid environmental damage would be contrary to those rights. This does not, however, mean that trade-offs disappear. Rather, the specific vestment of rights merely changes the type of trade-off that occurs in any choice and hence changes the type of stated preference question that can be asked. Instead of asking for payment to avoid environmental loss, the question has to be worded around the notion of compensation. Rather than willingness to pay, the value concept switches to willingness to accept compensation.

It can't be done

The second dimension of the challenges facing valuation without market prices involves technical difficulties in applying stated preference surveys. There are those who say that stated preference surveys can't be done because they are likely to produce biased estimates of environmental values. This is because they are not based on the revelations of preference strength that take place in markets.

This is more of a technical argument and it relates to the incentives for survey respondents to tell the truth when answering questions

---

8   The operations of Nature Conservancy (http://www.nature.org/) and Bush Heritage Australia (http://www.bushheritage.org.au/) illustrate this behaviour.

that are necessarily set in hypothetical contexts.[9] In stated preference surveys, people are asked to make trade-offs that do not currently exist. Respondents are asked to put themselves into hypothetical situations in which they are able to pay (perhaps as an additional tax payment to government) to achieve environmental improvements or avoid environmental harm. Given that no money changes hands and that respondents' identities are kept confidential, respondents may be tempted to give answers that they think will influence the policy choice in a way that will benefit them.

Hence, in the earliest stated preference surveys when respondents were asked the maximum amount they would be prepared to pay for an environmental improvement, the incentive for those who held positive values for the environment was to over-state their true willingness to pay. By doing that, they hoped to secure the improvement and not have to pay the amount they stated.

Alternatively, if respondents think that the questions they are being asked are purely hypothetical, they may decide it isn't worth the effort of thinking carefully about their preferences. The outcome then would be willingness-to-pay statements in survey responses that are unrelated to real preferences.

These technical issues have been the subject of a great deal of research effort over the last 30 years. The result has been the development of refined techniques designed specifically to address the issue of survey response bias. This research effort has not been limited to environmental contexts, as the same challenges have faced decision-making in public health and transport applications. Even commercial marketing research has focused on the development of stated preference techniques. Businesses seeking to launch new products need to explore their prospective demand. Because the products are not yet marketed,

---

9   The argument may extend beyond the technical aspect to a broader concern that values only emerge through action. If this is accepted, hypothetical situations that involve no action cannot, in principle, enable the revelation of values.

questions posed to survey respondents about their likely purchasing behaviour are hypothetical. Getting accurate answers about preferences in such cases is a key component to the commercial success of a product launch.

One development arising from the stated preference research agenda has been Choice Modelling (CM).[10] In this type of survey, respondents are asked a sequence of albeit hypothetical choices between different future options. Because the options vary in what they offer, it's possible to see from people's responses how they trade off between the various characteristics of the options. By using characteristics that are a mixture of environmental features (such as the area of wetlands protected or the concentration of pollutants in the air) and money (perhaps an additional tax or a higher utility price), the trade-off between money and the environment can be observed.

Choice Modelling practitioners aim to convince their survey respondents that the context of the questions being asked is real and that the answers they provide to the choice questions will have consequences for decision-making. That helps to avoid the prospect of people not taking care when answering. The sequence of questions asked in a CM survey also makes it difficult for respondents to work out how they could manipulate their answers strategically.

And it is done

With the advances made in stated preference survey design and application, more and more studies are being performed to inform government policy. Three recent and high-profile decisions in Australia have involved environmental value information generated through CM applications.

Decisions made by the Victorian Government in Australia to establish new national parks and nature reserves along the River Murray

---

10 Bennett, J. and R. Blamey (2001), *The Choice Modelling Approach to Environmental Valuation*, Edward Elgar, Cheltenham.

to protect the river red gum forests were informed by a benefit–cost analysis of alternative policy initiatives relative to a do-nothing-new option. The environmental benefits of the proposed parks were estimated using a choice modelling study.[11] Respondents were asked to trade off between forest management options that included more protected forest, increased recreation facilities and higher levels of taxes and charges.

An environmental impact assessment prepared as part of an application to extend the underground coal mining operations at the Metropolitan Mine, near Sydney, included a benefit–cost analysis. The costs of environmental damage projected to be caused by mining – mostly relating to the cracking of stream bed rocks caused by subsidence – were estimated using CM. The assessment was used by the state government planning department in their determination to allow mining but with restrictions to limit the extent of stream damage.[12]

A proposal to introduce a national scheme for recycling end-of-life computers and televisions was supported by a Regulatory Impact Statement that included an estimation of the environmental benefits of the scheme derived from a CM application. Samples of people from the major urban centres were asked to choose between alternative recycling schemes and an option where no new scheme was introduced. The choices were characterised by different percentages of recycling achieved and different collection schemes. Each option apart from the no-new-scheme option came at a cost: To achieve more recycling required an additional payment. The Regulatory Impact Statement was used by state and federal environment ministers in the policy debate that led to the introduction of the recycling scheme.[13]

---

11 http://www.veac.vic.gov.au/documents/VEAC_Final_CM_report_1_June_07.pdf
12 http://www.peabodyenergy.com.au/nsw/metropolitan-coal-project-environmental-assessment-2.html
13 http://www.ephc.gov.au/taxonomy/term/51

The use of stated preference surveys to inform policy decision-makers about the trade-offs people are prepared to make is not only an Australian phenomenon. For example, a choice modelling study investigating the values of landscape features in the upland regions of England was used as an input into the re-design of assistance payments made to farmers in 'Severely Disadvantaged Areas' so that they reflected environmental characteristics differences.[14] And in the United Sates, a wide range of policy determinations and cases at law have been informed by stated preference studies.[15] Of particular note are the compensation determinations made under the 'Superfund' legislation whereby pollution damages can be recovered by state and federal governments.[16] An inventory of environmental valuation studies – the Environmental Valuation Reference Inventory (EVRI)[17] – has been established by the Canadian Department of the Environment with the financial support of the UK, US, French, Australian and New Zealand Governments to facilitate the use of these values across different policy applications.

A 'little green lie'?

If the environment were infinitely valuable, humanity would not be able to exist. Our very existence causes harm to be done to the environment. That harm would not be tenable if the environment had 'trump status' over all other values.

Because we choose to continue our existence, the proposition that the environment is infinitely valuable is flawed. Why, then, do people profess the infinite value proposition and refuse to accept the prospect of any environmental harm, even when their own day-to-day actions

---

14 Hanley N., S. Colombo, P. Mason and H. Johns (2007), 'The Reform of Support Mechanisms for Upland Farming: Paying for Public Goods in the Severely Disadvantaged Areas of England', *Journal of Agricultural Economics*, **58** (3), 433–453.
15 Thomas H. Stevens (2005), 'Can Stated Preference Valuations Help Improve Environmental Decision Making?', *Choices*, **20** (3), 189–193.
16 http://www.epa.gov/superfund/index.htm
17 https://www.evri.ca/Global/Splash.aspx

have negative environmental impacts? Calls for environmentally harmful activities to be banned remain a prominent feature of many environmental lobby groups' campaigns. For example, coal-fired power stations are a frequent target.[18] But so too are overhead power lines that are built to transmit electricity from wind energy 'farms' located in scenically attractive and relatively remote locations.[19]

One explanation of this apparent conundrum is that people adopt the extreme position as a bargaining strategy. In the realm of public debate, requiring an activity to be banned may provide a starting point for negotiations in which trade-offs are then considered.

But there is also the explanation that the position is taken by environmentalists and their lobby groups without any thought of 'backing off'. The idea would be to achieve an environmental goal for which they have strong preferences and high values without having to bear the costs. If the costs of securing an environmental 'ban' are imposed on others through a policy determination, then it is worth pursuing. Why make a trade-off if someone else can be made to bear the costs and you can retain the benefits? So if a ban on coal-fired power stations ensures that your environmental goal is achieved and the costs of that ban are met by others (especially large electricity consumers), then the policy provides you with strong gains and few costs. It makes sense to lobby for the ban. However, it may not be the most appropriate strategy from a community-wide perspective where the environmental gains must be weighed up against the power supply costs.

In that regard it is not surprising that environmental groups are frequent critics of the use of stated preference techniques to estimate the extent of environmental costs and benefits. If the estimation of these otherwise intangible values clarifies the trade-offs involved in public-sector decisions, then there is a risk they may show the environmental benefits to be less than the resource development benefits. In that case,

---

18 http://www.ft.com/intl/cms/s/0/6f775570-99c4-11df-a0a5-00144feab49a.html
19 http://www.stirlingbeforepylons.org/

the environmental lobbyist would rather not have the non-priced values quantified. It would be preferable to be able to keep the debate limited to qualitative appreciation of the environmental impacts. Degrees of negotiating freedom are thus preserved. The same applies to advocates of resource development when the benefit–cost analysis shows the environmental costs to be greater than the development benefits.

This makes supporting benefit–cost analysis a risky business for lobbyists on both sides of the development/protection divide: The answer provided by such an analysis may not be supportive of their desired outcome.

But it may also be difficult to find friends for benefit–cost analysis and inherently, stated preference environmental valuation, amongst the policy-making/advising community. Politicians' goals usually include re-election high on the list. Being able to make decisions so as to elicit political support is therefore a valued part of the politician's duties. Having benefit–cost analyses to inform decision-making reduces decision-making flexibility. If a decision is taken that contravenes the findings of a benefit–cost analysis, the politician responsible will have to make greater effort to defend his or her decision. Hence, the opportunities to be able to make decisions that give favour to members of the electorate who are important to re-election prospects are restricted.

With this weight pulling against the use of benefit–cost analysis and stated preference techniques for environmental valuation, it is not surprising that they attract so much criticism. Little wonder also that alternative techniques such as multi-criteria analysis (outlined in Chapter 3), which can be manipulated in order to produce recommendations that suit the various interest groups and political needs, are so lauded despite their fundamental flaws.[20]

In this light it is also no surprise that some respondents to stated preference surveys have problems answering the sort of questions put.

---

20 Dobes L. and J. Bennett (2009), 'Multi Criteria Analysis: Good enough for government work?', *Agenda*, **16** (3), 7–30.

In a choice modelling survey, people are asked to make trade-offs when choosing between alternative futures. In doing so they are forced to confront the TANSTAAFL principle (There Ain't No Such Thing As A Free Lunch): If they want to have an improved environment, they have to give up something. It is entirely expected that people would like to have goods and services, including the environment, without any cost. Stated preference questions can make people uncomfortable when the option of having more of everything ceases to be available.

The questions also force people to recognise the presence of a budget constraint. Not being able to pay as much as you might like because of a limited budget can be frustrating and it is always a good prospect for people to look for others to pay for what they want but can't afford. 'Others' in many cases can be defined as 'the government'.

All of these points demonstrate the risks involved when resource-use decision-making is handed from individuals trading in markets to governments acting as the 'collective'. The prime danger is that in taking on the collective responsibility, governments – both politicians and civil servants – do not act in the best interests of the collective but rather pursue their own personal best interests. The classical 'principal–agent' problem emerges whereby the principals (citizens) are unable to ensure that their agents (politicians and civil servants) act on their behalf in their best interest.

Using stated preference study results in benefit–cost analysis for evaluating environmental policy options is an important component of any strategy designed to overcome the principal–agent problem. Making sure that the public is aware of the consequences of public policy decisions and that people's values (for environmental and other consequences) are incorporated into the evaluation process are vital. But so too are political processes that ensure citizens can respond to poor decision-making.

9: Climate Change

Proposition: We must reduce greenhouse gas (GHG) emissions to avoid global climate change.

Human-induced global climate change is a serious threat to the continued ability of the planet to support humanity and current ecosystems. The damage caused by climate change will be so large that GHG emissions must be reduced now.

BUT

Reducing GHG emissions would be costly. The decision to bear those costs should be made with reference to the expected benefits that reduced GHG emissions would provide. Reducing GHG emissions will not eliminate the risk of climate change.

A hot debate

Perhaps the most vexed environmental debate of the current era has focused on the issue of climate change. If it is not the prime focus of environmental policy initiatives, it seems that climate change is used as a justification for a host of other measures relating to different policy issues. Measures aimed at reducing greenhouse gas emissions, such as carbon taxes and trading schemes, have been introduced in the European Union, Japan and New Zealand and proposed in numerous other jurisdictions including Australia, California and Canada. These are the mitigation (or abatement) policies aimed at reducing the impacts of climate change.

Other environmental policies are aimed at adapting to future climate change. These are as diverse as those relating to biodiversity protection and invasive species control. And the impact of, and adaptation to, climate change extends well beyond environmental policy making. Under future climate scenarios, it is argued that patterns of health expenditures will have to be altered: for instance the range of malaria incidence is expected to increase and heat stress amongst elderly people is also anticipated to become more prevalent. The advent of climate 'refugees' attempting to escape from countries adversely affected by sea level rises or increased drought incidence and water shortages is anticipated to require changes to immigration policy and even more border defence spending.

Even the topics specified by research funding agencies as being of highest priority have seen climate change-related themes promoted to the top of the list.

The ramifications of climate change are portrayed as being monumental for societies around the globe. It is also apparent that the impacts of both climate change mitigation policies and policies relating to climate change adaptation are already significant and, potentially, also monumental. Australian Treasury estimates indicate that a carbon tax would cost the Australian economy the equivalent of one-and-a-half years' national income in the period to 2050. That cost rises to three years' national income if those nations with which Australia competes on international markets do not impose a similar tax regime.[1]

Not surprisingly, then, the climate change debate is both extensive and at times bitter and acrimonious. The stakes are high for those who believe that climate change will affect the future capacity of humanity to survive. They are also high for those whose livelihoods depend on the continued success of industries that are greenhouse gas emitters and for consumers who enjoy the products of those industries. Given the current

---

1   http://www.theaustralian.com.au/national-affairs/commentary/climate-policy-a-burning-issue/story-e6frgd0x-1226072611617

dependence on fossil fuels in the production and transportation of almost everything we buy, the latter 'consumer' category is particularly broad-based.

But there any many others whose well-being is involved: researchers in the science and policy of climate changes, policy makers as well as their advisors, bureaucrats responsible for the administration of the mitigation and adaptation programmes implemented and those who do or would benefit from climate change policy measures. The latter group is potentially large and extends from financial market traders (who would become active in carbon trading schemes) through to renewable energy suppliers (whose products become more competitive because of climate change-inspired government interventions).

One reason for the climate change debate being so divisive, beyond the extent of the stakes involved, is that the issues involved are both complex and encompass numerous areas of expertise, none of which any single person is likely to have the capacity to understand completely.

People therefore come to the debate with differing vested interests and differing capacities to comprehend any or all of the complexities involved. And the inputs of populist offerings such as former US Vice-President Al Gore's movie 'An Inconvenient Truth',[2] the apocalyptic 'The Day After Tomorrow'[3] along with 'The Great Global Warming Swindle'[4] do little to clarify the matter in the eyes of the general public. Nor do the almost constant claims that the current weather patterns (hottest day, driest month, wettest year, most snow falls), or even the weather patterns over the last few decades (most/least cyclones, droughts, floods), prove or disprove the climate change hypothesis.

The science
The first element of the divergence in views regarding climate change

---

2 http://www.aninconvenienttruth.com.au/truth/
3 http://www.imdb.com/title/tt0319262/
4 http://www.greatglobalwarmingswindle.co.uk/

is in the physical science. In this field, two fundamental questions require answers: Is the world's climate changing? If so, is humanity responsible for any changes that are occurring? The findings of the Intergovernmental Panel on Climate Change (IPCC)[5] suggest that the answer to both of these questions is 'yes'.[6] The IPCC was set up by the World Meteorological Organization (WMO) and the United Nations Environment Program (UNEP) in 1988 to review available scientific studies into climate change. It has produced a series of Assessment Reports, the second of which was a key input to the negotiations that led to the establishment of the Kyoto Protocol for greenhouse gas emission controls in 1997. Under the terms of the Protocol, signatory nations agreed either to limit their greenhouse gas emissions to negotiated levels or to buy 'carbon credits' generated from mitigation activities undertaken in developing countries.

It is asserted by the IPCC (and others seeking to support policies that aim to reduce greenhouse gas emissions) that its position on climate change represents a 'consensus' amongst the scientific community.

Yet there is little doubt about the extent and strength of opposition to the IPCC's position on at least the second of the two questions:

---

5   The IPCC 'reviews and assesses the most recent scientific, technical and socio-economic information produced worldwide relevant to the understanding of climate change. It does not conduct any research nor does it monitor climate related data or parameters. Thousands of scientists from all over the world contribute to the work of the IPCC on a voluntary basis. Review is an essential part of the IPCC process, to ensure an objective and complete assessment of current information. IPCC aims to reflect a range of views and expertise.' (See: http://www.ipcc.ch/organization/organization.shtml). It is important to recognise that the IPCC is an 'intergovernmental' body. It is the member governments, therefore, that decide on the work programme undertaken, participate in the review process and make decisions regarding the acceptance, adoption and approval of the reports.

6   http://www.ipcc.ch/publications_and_data/publications_ipcc_fourth_assessment_report_synthesis_report.htm

Numerous so-called 'sceptics'[7] assert that climate change is not being caused by humanity. Some base their deviation from the IPCC 'consensus' on observations regarding past climatic variation across geological time, arguing that past climatic variations have not been correlated with levels of atmospheric carbon dioxide. Put simply, the contrarian position is that climate change (sometimes sudden and catastrophic) has been a feature of geological history but that the greenhouse gas emissions produced by people's actions are a minor and insignificant part of the process that determines climatic shifts.[8]

Given this level of dissent from the IPCC position[9] it would seem presumptuous to accept the claim of consensus. Indeed it would seem contrary to sound scientific method to accept that any hypothesis can ever be proven absolutely.[10] Rather, maintaining human-induced climate change as an hypothesis that may be accepted or rejected upon the production of evidence leaves open the opportunity to challenge the orthodoxy of the IPCC position. Closing off that opportunity runs the risk of preventing further evidence from being considered and prevents advances in our understanding of the forces at work in driving climate change from being attained.

At best, then, the IPCC position must be regarded as contestable rather than a consensus: There is sufficient dissent from enough sufficiently well qualified scientists to maintain a robust questioning of the human induced or 'anthropogenic' climate change hypothesis. Geological observation provides evidence of climatic variability

---

7   Including Czech Republic President Vaclav Klaus (http://www.klaus.cz/english-pages/) and Christopher Monkton of Brenchley (http://scienceandpublicpolicy.org/monckton/)
8   Plimer, I. (2009), *Heaven and Earth. Global Warming: The Missing Science*, Connor Court Publishing, Melbourne.
9   As represented in a series of four International Conferences on Climate Change specifically designed to challenge the IPCC's claims of consensus: http://www.heartland.org/events/2010Chicago/
10 Nor should 'truth' be determined by majority vote.

across past millennia and there is nothing to suggest that this will not continue into the future. However, it must be concluded that there is some chance that human actions are causing climate change and some chance that they are not. This contestability is the stuff of scientific endeavour. Continued experimentation and testing will give rise to a more refined understanding of the complexities involved in the cause–effect relationship between greenhouse gas emissions and climate. With greater knowledge so generated, the more (or less) confidence we can hold for the hypothesis. But it is essential that the hypothesis is not considered proven given that there is always the opportunity for further testing which may lead to it being rejected.

Just exactly what the chances (that is, probabilities) are of climate change being caused by humanity have not been addressed. The IPCC reports and other studies[11] deliver assessments of potential future temperatures as probability distributions: In other words, they provide a range of possible future scenarios of global average temperature at a point of time in the future along with the probabilities of each temperature occurring. However, the range of possibilities used by the IPCC does not include the assessments of those who dissent from the IPCC position. The implication of this omission is that the range of possibilities and their associated probabilities of occurrence has not been analysed.

The policy

The prospect of humanity causing climate change implies that humanity also has the power to change its behaviour so as to reduce its impact on climate. This is an important element in the consideration of any policy approach to mitigating climate change. It means that if it can be established that the expected benefits arising from human

---

11 Including those of the CSIRO (http://www.publish.csiro.au/?act=view_file&file_id=CSIRO_CC_Chapter%203.pdf)

action to mitigate climate change exceed the costs of doing so, then it is worthwhile for society to pursue that action.

For such an assessment of any climate change mitigation policy, it is necessary for science to provide extra information beyond answering the question: 'Do people cause climate change?' Specifically, the relationships between human actions, climatic conditions and the impacts of climate need to be estimated.

It is necessary to be able to predict what will happen to global average temperatures in the future if no policy action is taken to reduce greenhouse gas emissions. Then predictions regarding future climatic conditions must be made for a range of alternative greenhouse gas reduction policy options. In this way, climatic conditions 'with and without' policy initiatives can be predicted and so the change in climatic conditions created by the policy can be recognised. Then, the impacts of the predicted change on things that people value need to be estimated.

Hence, scientific analysis is needed to estimate the change in climatic conditions brought about by a policy. But it is also required to predict resultant changes in things as diverse as food production, the state of human health, weather-related property damage and the fate of endangered ecological communities.

None of these estimates and predictions can be made with certainty, so the scientific analysis must include the probabilities associated with possible outcomes.

Given the extensive nature of the impacts of climate, the time periods involved and the inherent complexities of the systems that are integral to the formation of the world's climatic conditions, the analytical burden imposed by this approach to policy assessment is considerable. Such complete information is not currently and never will be available. This, however, should not be a barrier to the consideration of policy initiatives. Decisions are constantly being made about the future in the absence of perfect information. However, the logic of the assessment process and the limited amounts of information should still be applied to the policy choices being made.

Policy assessments to date

A number of major reports on the policy implications of climate change have been produced. In the UK, the Stern Review[12] was commissioned by the Blair Government to recommend policy responses to climate change. In Australia, the Garnaut Review[13] provided input to the Rudd Government's policy formulation.

Both of these extensive studies follow the same logic in reaching their common conclusion in support of government action to reduce greenhouse gas emissions: the costs of taking policy action to mitigate climate change are small compared with the costs that people will have to bear if no policy action is taken. The analyses provide detailed predictions on the costs of mitigation, necessarily making assumptions regarding the capacity and willingness of people to substitute out of high greenhouse gas emitting activities into 'low carbon' futures. They also use the IPCC predictions of climate change as their basis for estimating the likely damages to the economy and to key features of the environment. Because these costs occur at different times in the future, the time-value of money had to be integrated into the analysis using the discounting process.

The presumption made under this logic is that the benefits of implementing a policy are equal to the costs that would arise if no policy were implemented. This necessarily assumes that the policy measure taken will provide an immediate and complete reversal of any climate change that humanity has created. Put simply, the costs of inaction are taken to be equal to the benefits of action. This is clearly not the case, as any policy will have a considerable lead time in taking effect; such is

---

12 http://webarchive.nationalarchives.gov.uk/+/http://www.hm-treasury.gov.uk/sternreview_index.htm
13 www.garnautreview.org.au

the nature of climatic response. It is also unlikely that anything but the most draconian greenhouse gas reduction policy would be capable of ever completely reversing any change to climate caused by humanity. At best, any policy measure will have some (but not total) mitigating impact and that impact will occur over time and at some stage into the future.

The logic employed by both Stern and Garnaut therefore departs from the standard policy assessment process that is inherent in benefit–cost analysis and well accepted in economic theory and practice. Both studies therefore tend to exaggerate the benefit side of the equation. The benefits of a mitigation policy are bound to be less than the costs caused by climate change: Policy measures will never be 100 per cent effective or instantaneous.

The interest rate dilemma

Both studies also opt for an approach to discounting that departs from standard benefit–cost policy assessment. The discounting process recognises that people value benefits and costs that occur in the future less than if they were to occur right now. It reflects people's preferences for time in the same way as people's behaviour in borrowing and lending: To have something of value now rather than in the future, people are prepared to pay an interest rate to borrow and people who are willing to forgo access to current consumption need to be compensated through the receipt of interest payments. The rate of interest therefore reflects people's 'time value of money'.

Most benefit–cost assessments thus use market interest rates as the starting point for their discounting process. Adjustments are made to that rate to reflect the differing structure of risk facing governments (and hence communities) when making choices. These adjustments are usually toward lower rates. This is because communities can spread the risk across a wide range of different investment opportunities. It is also because the time frame of choices made by society is different from that of the individuals: While a person will borrow and lend in money markets knowing that they face a limited life span, a whole community comprises a sequence of generations without a finite end-point.

However, even with these types of downward adjustments, the interest rates recommended by most government finance departments are around the five per cent level in 'real' terms.[14] Anything less than that will be a cause for concern because it can cause 'crowding out': with interest rates applying to the public sector being so much lower than the private sector, more investment projects in the public sector will pass the test of generating a rate of return greater than the borrowing rate than in the private sector. This provides a signal to divert resources (through higher taxes) to government control.

With a five per cent interest rate, the discounted value of benefits occurring beyond 50 years into the future becomes very small. $100 received in 50 years' time has a present value of $7.69. If it is received in 100 years' time, it is worth just 59 cents now. Because the benefits of any climate change policy introduced now would not be enjoyed until well into the future, while the costs start to be borne immediately, the standard discounting process makes almost any mitigation policy fail the benefit–cost test.

The Stern and Garnaut reports use an interest rate that approaches zero for their discounting calculations on the grounds that the interests of future generations must be given special consideration when climate change policies are being considered. This is acknowledged as a 'normative' approach: one that is based on the views of the analysts as being most appropriate. This is in contrast to a 'positive' approach that is based on observations of people's behaviours in markets.

Using lower-than-market interest rates as a mechanism for supporting 'conservation' investments that are initially costly and beneficial only over the longer term is a two-edged sword. If used for project evaluation

---

14 That is, adjusted for the impact of inflation.

beyond climate change initiatives, lower rates not only promote 'farsighted' conservation investments and policies, but they also mean more investment projects in general pass the benefit–cost test. That means faster use of resources in the near term. It also means that fewer people have an incentive to save for the future. Neither of these side-effects is generally welcomed by conservationists.

Another impact of using a low interest rate is that it fails to recognise the true opportunity costs implied by time. The market interest rate demonstrates that resources can be used now to generate a stream of well-being over time. Taking them out of those market-directed uses and allocating them to use in climate change policies that imply much lower interest returns means a loss of well-being. In other words, the implementation of climate change mitigation policies, as advised by Stern and Garnaut, will mean that losses of well-being will be endured by the current generation and by future generations who will not 'inherit' as much wealth from the current generation as would otherwise be the case. That wealth could be in the form of the manufactured capital stock (such as roads, ports and factories) but also in human capital form (health and education levels in the population) and even in natural capital (improved air and water quality and nature conservation areas).

In both the definition of benefits and the application of the discounting process, the Stern and Garnaut assessments of climate change policy options departed from conventional benefit–cost analysis principles. Both departures systematically favoured the adoption of mitigation policies. The treatment of mitigation policy benefits provides an illustration of this.

Mitigation policy benefits

The benefits of adopting a mitigation policy are the costs of climate change that would be avoided because of the policy. The first step in understanding this value is predicting the effectiveness of the policy:

Will the mitigation of greenhouse gas emissions achieved by the policy have an impact on climate? Following that, the extent of the effects of the change in climate achieved on the values held by people must be estimated.

Information on both of these steps is lacking. However there is some knowledge of the likely effect of mitigation on climate. For instance, Lomborg[15] finds that complete international compliance with the Kyoto Protocol for greenhouse gas emission reductions would have the effect of delaying the impacts of climate change by four years after 100 years. The benefit of the mitigation effort sought by the Protocol is therefore the postponement of the costs of climate change rather than an avoidance of them. An implication of these findings is that the benefits of climate change mitigation policies are likely to be much smaller than the costs of climate change.[16] Mitigation policies would appear to make only minor differences to the projections of future temperatures that are made by the IPCC under the 'business as usual' scenario. Given that those changes will occur well into the future, the impact of standard discounting will have the effect of further reducing these anticipated benefits relative to the estimates advanced by the Stern and Garnaut reports.

The benefits of most mitigation policies estimated using the appropriate benefit–cost analysis framework are so small that they are likely to be dwarfed by the policy costs. To have an effect on the climate that is beyond the trivial and in a time frame that limits the effects of discounting, mitigation policies would need to be so extreme that the

---

15 Lomborg, Bjørn (2007), *Cool It*, Random House, New York.
16 There are also off-setting benefits associated with climate change. For instance, the agricultural production of some regions is likely to rise and deaths associated with cold winter temperatures could decline. Furthermore, when uncertainties regarding the effectiveness of policy instruments such as the Kyoto Protocol to deliver GHG emission reductions and then for those reductions to have an impact on climate are added into the estimation of the *expected* benefit of mitigation effort, the relative costs of effort become even greater.

costs would be so large as to make them politically unattractive as well as still being greater than the albeit enhanced benefits.

Free riders?

The assessments of climate change mitigation policies carried out in the Stern and Garnaut reports were both commissioned by individual national governments and were aimed at providing policy advice to their respective governments. However, climate change is a global rather than a national issue and national policy measures need to be considered in the light of international action.

Indeed much of the debate focusing on climate change has been to do with the process of seeking international agreement on the target level of greenhouse gas emissions reductions each nation is willing to make.

The Kyoto Protocol involved developed and 'economies in transition' (EIT)[17] countries[18] committing to reduce their emissions. Each signatory agreed to a separate target reduction and in aggregate the emission level target was five per cent below the 1990 level during the first commitment period from 2008–2012. Part of the reductions in emissions could be sourced from the purchase of 'carbon credits' from developing country signatories that reduce their emissions or that sequester carbon through activities such as reforestation.

Clearly there were incentives for the developing countries to commit to the Protocol. Theirs was a no-loss commitment: They could make money from carbon reduction schemes. The situation was different for most Annex I countries as a commitment was (for most) to involve costs. The extent of those costs varied according to the amounts of the cuts negotiated and their capacities to reduce emissions. For some countries the burden was light or non-existent. For instance, countries in the former Soviet Union such as Russia and Ukraine agreed to

---

17 Predominantly those nations comprising the former Soviet bloc.
18 Referred to as 'Annex I nations' by the United Nations Framework Convention on Climate Change – the UNFCCC.

hold their emissions constant. Such was the extent of the industrial decline in those countries since the base year of 1990 that they ended up being well within their commitment levels.[19] Australia managed to negotiate a light burden, with 108 per cent of 1990 emission levels being the 2008-2012 target because of land-clearing activities taking place in 1990. For other countries, notably the USA at 93 per cent, the commitment requirement was more costly to meet. The USA refused to sign the Kyoto Protocol.

Progress towards meeting the agreed levels of emissions has been mixed. By 2008, all developed nations had reduced aggregate greenhouse gas emissions (excluding emissions/removals from land use, land-use change and forestry) to six per cent below the 1990 base level. The five per cent target seems to have been achieved. However, much of this reduction was achieved through the de-industrialisation contributions of the EIT nations: The emissions of these countries declined by 36.7 per cent over the period. The global recession that occurred in 2008 also had a negative impact on emissions as economies experienced reductions in economic activity. The non-EIT Annex B parties to the Kyoto Protocol increased their emissions by eight per cent between 1990 and 2008. For example, Australia increased its emissions by 31 per cent (compared to its eight per cent increase target), Canada by 24 per cent (its target was a six per cent reduction), and Turkey almost doubled its emissions. Emissions from the United States increased by 13 per cent.[20] Some European Annex I nations were able to reduce their emissions by more than the aggregate target of five per cent: France managed a 5.9 per cent reduction (which is lower than its

---

19 This is the origin of the so-called 'hot-air' issue. Under the Kyoto Protocol, countries were granted a certain number of permits to release greenhouse gases on the basis of 1990 levels of emissions. With the de-industrialisation that followed the collapse of the USSR, the surplus permits (hot-air) threatened to flood the EU carbon market (http://www.pointcarbon.com/).
20 United Nations Framework Convention on Climate Change (2011), *Compilation and Synthesis of Fifth National Communications*, FCCC/SBI/2011/INF.1, Bonn.

nation-specific target of eight per cent) and the UK achieved 18.5 per cent (against a target of eight per cent, largely because of a shift from coal to gas electricity generation over the period). These performances were enhanced through the trend for carbon-intensive industries to relocate from Europe to Asia.

The costs of reducing emissions have similarly varied across countries. The Australian Productivity Commission found that current spending on carbon abatement in Australia is between $44 and $99 per tonne in the electricity generation sector, whereas Germany spends $137–$175, the United Kingdom $75–$198, the US $43–$50 and China $35–57 per tonne. Solar power systems were costly means of emission reduction when compared with other technologies, with around $137 being paid to prevent one tonne of carbon emissions. Most biofuels policies were also found to be relatively high-cost means of achieving abatement, with US subsidies on the production of ethanol amounting to $672 per tonne of carbon.[21]

The differences between country performances in reducing emissions and the spread of costs help to explain why negotiations to establish an agreement to follow on from the Kyoto Protocol have stalled in Conferences of Parties (COP) held in Copenhagen, Cancun and Durban. The larger the numbers and the more divergent the interests of those countries involved, the less are the chances of securing international agreement. When interests do not readily align, the temptation for 'free-riding' behaviour comes to the fore. Individual nations have the incentive to offer little if any commitment in the hope that others will commit sufficiently to see the risks of potential climate change reduced without any costs to themselves. Furthermore, developing countries, which tend to have the strongest carbon growth trajectories as their economies transform, argue that their carbon emissions should not be curtailed given that developed countries have

---

21 http://www.pc.gov.au/projects/study/carbon-prices/report

had their past growth stimulated by having free access to the atmosphere as a 'waste dump' for their carbon emissions.[22]

Without such a multi-lateral agreement in place, individual nations must make policy decisions regarding their greenhouse gas emissions given the chance that other nations will not take any mitigation action. In that case, while the costs remain the same, the benefits of a policy initiative fade even further: Reductions in the emissions of one country (even a large one) will by itself have little if any impact on global climatic conditions.

The politics of it all

The political feasibility and economic efficiency of policy measures are also functions of the values held for policy change by individual voters/citizens. The paramount question in both contexts is whether or not the public is willing to support a political party that proposes to introduce a climate change mitigation policy and whether or not they are willing to pay the costs involved.

To date, there has been insufficient political support for an economy-wide carbon tax/permit scheme to be introduced in the USA, Canada or Australia. This is despite, at least in Australia, a number of high-profile government-funded publicity campaigns in support of climate change-related policies. Levels of support for the introduction of a carbon tax in Australia fell to around 38 per cent of eligible voters in June 2011.[23]

A key element of people's willingness to support the introduction of a tax or other greenhouse gas abatement scheme is the extent of the costs they will have to bear. Individuals will weigh up the benefits they

---

22 See Brennan, G. (2010), 'Climate Hopes: Pious and otherwise', *Australian Journal of Agricultural and Resource Economics*, **54** (1): 5–7.
23 http://www.theage.com.au/environment/climate-change/carbon-cate-is-just-fodder-for-coalition-campaign-20110531-1fdlf.html. Subsequently, the Australian Parliament passed legislation to introduce a carbon tax but the opposition party has pledged to repeal the legislation if elected.

expect to enjoy from the introduction of a policy against the anticipated costs. For a policy aimed at addressing climate change, it is likely that people will consider the benefits not only to themselves directly but also to future generations in this decision. One way to investigate the extent of the benefits people expect to enjoy from alternative policies is to see just how much by way of costs they would be willing to pay in order to have a policy implemented.

The results of a study (conducted in 2009) that asked such questions of a sample of Sydney residents about the introduction of an emissions trading scheme in Australia found an average willingness to pay of AUD58 per household per month.[24] This is around three per cent of the sample's average household income. This amount reflects the benefits that survey respondents expect to enjoy from the avoidance of climate change impacts offered by the proposed scheme. It should be compared to the estimated costs that the scheme would impose on households. These have been estimated at AUD624 per annum (or AUD52 per month).[25] Does this mean that the policy would generate a net benefit to society given a cost of AUD52 and a benefit of AUD58 per month? That depends on how well the sample represented the Sydney population. While the study shows that the socio-economic characteristics of the sample closely align with those of the Sydney population, not all who were asked to respond to the internet-based survey chose to do so. That implies there was a fraction of the sample that may not be willing to pay anything to have an emissions trading scheme introduced. With sample response rates that fall between one-third and a half, the benefit–cost equation indicates an overall lack of

---

24 Akter, S. and J. Bennett (2011), 'Household perceptions of climate change and preferences for mitigation action: the case of the Carbon Pollution Reduction Scheme in Australia', *Climatic Change*, **109** (3-4): 417–436.

25 http://www.climatechange.gov.au/government/initiatives/cprs/carbon-price-design/household-assistance.aspx.This estimate has been challenged by numerous analysts as a serious under-estimate (for example see: http://www.crikey.com.au/2009/11/09/the-real-costs-of-rudds-cprs-are-just-starting-to-surface/).

support. This result is substantiated by more recent polls that report support levels of around 40 per cent in favour of mitigation action.

A 'little green lie'?

The global warming issue involves many people with strong vested interests. It is no surprise that these people will view the issue from the angle that best suits their own case. And with so much uncertainty surrounding long-range climate projections (given the complexities of the process that forms climatic patterns) let alone the social, environmental and economic impacts of the potential climate outcomes, there is a wealth of alternative angles from which to choose. Hence what may be regarded as a 'little green (house gas) lie' by one party will be seen as the absolute truth by another.

It is therefore useful in trying to unravel the climate change policy issue to consider the likely winners and losers who would emerge from the introduction of measures designed to mitigate greenhouse gas emissions.

First there are the direct interest groups. These include those who are fearful of the prospects of climate change and would embrace those with strong preferences for the assets they see as being at risk in a changed climate, such as endangered ecosystems or low-lying waterfront property. Then there are the people who are most dependent on production or consumption processes that have a heavy carbon 'footprint': For example, coal mine owners and workers and those involved in the aluminium industry (as a heavy consumer of electricity).

Both of these groups will engage analysts in support of their claims. Once the analysts have established their cases, they will be inclined to maintain those positions. The continued policy debate is certainly in the analysts' best interest as it means return business.

Next, there are the indirect beneficiaries. These people will see potential gains from the introduction of the new policy. For instance, a carbon trading scheme will mean the emergence of a whole new set of financial gains from facilitating trades in the newly established market. Some brokers, bankers, insurance agents, economists and lawyers will identify these opportunities and will therefore be encouraged to support the pro-climate change interpretation of the world.

Also included in this group would be the bureaucrats responsible for researching then establishing the legislation, its subsequent administration including the international agreements entered into, as well as the monitoring and the enforcement roles. Already, the climate change issue has spawned an array of bureaucrats at every level of government (from local city/county 'climate change officers' through to federal ministry staff and, of course, UN officials). As well as financial gains, people in this group will enjoy heightened 'hierarchical importance' within and outside their organisation, have better chance of promotion as well as the chance to attend high-profile conferences and meetings in desirable international locations. And making sure that the policy structure is so complex that it is not only difficult for all to understand, but also tricky to administer, is also in the best interest of bureaucrats seeking to maintain and enhance their positions.

Politicians also have the potential to gain through the introduction of carbon abatement policies. For a start, these policies offer the opportunity to create a new source of tax revenue. With more tax revenue, members of the government have increased opportunities to redistribute wealth in the economy to advantage their supporters and to encourage opposition supporters to shift their voting allegiances. They can also use the additional takings to placate particularly vocal interest groups. For example, compensation payments to 'trade-exposed industries' may be promised. Low-income households may be given rebates for higher electricity prices. Car drivers may have their fuel costs subsidised. While all of these offsetting measures counteract the carbon reduction properties of a tax or emission trading scheme, they do enable politicians to at least partially neutralise any voter swing against the original policy. Put simply, the revenues generated can be used to consolidate political power.

From a political perspective, there are, therefore, a number of groups for which carbon abatement policies will have sufficient impact to shift voting intentions. Most of the costs of a carbon tax (or permit scheme) will, however, fall on members of the general public in a diversity of ways that, individually, may not even be noticed. The prices of most goods and services would rise, each by a relatively small amount: A change in the price of groceries due to higher energy input costs, or higher fuel and electricity prices, may be introduced over time in small increments so that many voters would be largely unaware of the policy's consequences. Or they may not be sufficiently affected to make a difference to their voting intentions.

This helps to explain why politicians in government are enthusiastic about introducing climate change policies and why they may be willing to engage in some 'little green lie' telling. And why they will receive whole-hearted support from their bureaucracies, the finance industry, environmentalists, and any businesses that can receive 'special treatment' to cope with the policy's impact that might be able to provide them with an edge over their competitors.

On the international stage, it is also conceivable that European nations, with their established specialisations in the use of manufactured capital, can see that an internationally applied carbon abatement scheme will retard the development of their emerging Asian competitors. These nations remain more reliant on cheap carbon-based energy sources for their production muscle and comparative advantage.

Despite the political push that carbon abatement policies can expect, it remains unclear that these policies are in the best interest of society, both at a national level and even at a global scale. While the costs of the policies are clear, apparent and immediate, the benefits are uncertain, poorly understood and only likely to occur far into the future.

On a national scale, a policy pursued on a unilateral basis is unlikely to have any impact on future climatic patterns. This is true for large-scale emitters such as the USA or China. It is glaringly obvious for small emitters such as New Zealand or even Australia with its heavy reliance on coal as an export commodity. Unilateral action will therefore not be worthwhile. Even multi-lateral action at a scale on a par with the Kyoto Protocol will produce such minor variation to the climate of the future that its benefits will be highly unlikely to exceed the costs.[26] And given the difficulties of negotiating such an international agreement, it could be expected that only very modest changes from the 'business as usual' scenario could be achieved.

That is unless cost competitive alternatives to carbon-based energy sources are developed. This has been demonstrated by the success of the Montreal Protocol on Substances that Reduce the Ozone Layer. The development of a cheaper substitute for chloro-fluoro-carbon (CFC) refrigerants gave producers an incentive to move away from the use of ozone-depleting substances.

Hence, if technological developments of substitutes were to be coupled with continued increases in the relative prices of fossil fuels, a shift to a low carbon economy will occur.

So if carbon emission mitigation/abatement policies are unlikely to deliver net benefits to society, should we continue 'business as usual'? The answer is 'probably not'.

Climate change poses a risk to society. That risk may or may not be due to human action, but it is a risk nonetheless. When confronted with the risk of a catastrophic outcome in the future, it is always worth contemplating taking out an insurance policy. In the climate context that can involve the adoption of adaptation policies. Essentially, these policies involve taking actions now that will protect society's interests

---

[26] Noting again that climate change is not universally damaging: Some areas of the world would become more habitable and productive in the event of a three degree increase in average temperature.

in the event of climate change causing a threat. For instance, if climate change is forecast to increase the risk of cyclones in an area, investments now in community shelter bunkers, enhanced sea walls and emergency power generators may be worthwhile. Greater incidence and severity of drought may warrant the construction of larger reservoirs and investment in breeding varieties of food crops that better withstand drought conditions. Provision of air-conditioned refuges in aged person facilities could save lives.[27]

Each of these adaptation policies would require careful assessment using benefit–cost analysis to ensure that net benefits were being created for society. Each one would require the use of scarce resources that could be used for other purposes, including educating the next generation, caring for the sick and frail, preventing the spread of infectious diseases, protecting rare and endangered ecosystems and assisting in the development of nations. Making the best choices between these competing uses of scarce resources remains a key priority for society. Guarding against the use of resources for policies that will deliver net losses to society – but would make some individuals better off and deliver electoral gains to politicians – is a task that can be aided by the application of benefit–cost assessments.

---

27 World-wide, more people die each year from the cold than for excessive heat.

10: Protecting the Environment, Privately

Proposition: The care of the environment cannot be entrusted to the private sector.

The environment provides 'public goods' that should be available to all free of charge. That means the government has to be responsible for caring for the environment. The private sector will either destroy it or try to profit from it.

BUT

The public sector will face problems managing the environment. Gathering information for effective decision-making is costly. Politicians' and bureaucrats' incentives can conflict with the public's best interest. Private solutions can be lower cost and better aligned with what the community wants.

Public goods?

To start off, a brief review of the 'conventional wisdom' regarding the nature of environmental goods and services is useful.

Many of the goods and services provided by the environment are characterised as being 'public goods'. This means that once the environmental asset is protected, the goods and services so produced are available to everyone, as it is either impossible or too costly to stop people from enjoying them. For instance, when an area of bushland is set aside as a protected conservation area, it provides a range of valuable goods and services: The water flowing from the area will be purified;

the genetic material of the area's plants and animals will be protected for future use; its scenic beauty will be maintained; and people will feel pleased because the risk of extinction facing the area's species will be reduced. Trying to stop anyone who hasn't paid to enjoy these benefits from enjoying them is tough.

What's more, once the conservation area has been established, it doesn't cost anything more to let more people enjoy the range of goods and services it provides. That means no extra scarce resources have to be used for more people to enjoy the benefits provided by the area. The implication of this is that all the goods and services provided by the conservation area should be available free of charge.

With no ability to prevent people from using these so-called 'public goods' provided by a conservation area, ownership issues arise. Property rights to 'public goods' are either difficult/impossible to define or defend. For instance, it may be possible to set up ownership rights over specific areas of land set aside for conservation. However, the rights to the clean water, the genetic material, the scenic vistas and the 'saving' species from extinction are all difficult to define, let alone defend. Those who haven't paid a contribution toward the purchase of the land for conservation can still enjoy the (public) goods and services it provides, free of charge.

Some of the benefits provided by a protected conservation area are enjoyed only by people who visit it. Passers-by may enjoy the scenic vista provided by the area, but those who enter the area will enjoy a different experience. But even this 'recreation benefit' provided by the conservation area can have some 'public good' characteristics. It may be technically possible to erect a fence around a protected area and only allow access to those who pay an entry fee. However, policing entry into an area for recreation may be so costly because of a 'porous' boundary and low numbers of visitors that it isn't feasible to do so. And if all of these goods can be provided to an extra user without imposing any extra costs, then anything other than charging a zero price would send the wrong signal. Some people wouldn't use the goods if they had to pay for them. But if it doesn't cost anything for them to use the goods, then all of the benefits they could have enjoyed from using the goods are lost.

Public action

Without the capacity to structure a set of property rights over these environmental public goods, 'ownership' cannot be established. Without ownership, there is nothing to buy and sell. That means markets won't be able to form and provide profit incentives for people and businesses to supply environmental protection. Although it may not cost anything to allow an extra person to enjoy these public good environmental benefits, it does cost something to set them up in the first place. For instance, establishing a national park will mean bearing the costs of not being able to use the resources that make up that national park in other profit-making ventures (such as timber harvesting or housing estates). With these costs to bear and no chance of earning any revenues because of the problems associated with stopping users who don't pay, private profit-motivated supply will be unlikely. And if the environmental asset exists but is un-owned, the incentive is for people to use it to the point where they can extract no more value from it. This is Hardin's 'tragedy of the commons':[1] If users can't be prevented from accessing a resource, it will be in danger of over-use. These 'open-access' resources can include agricultural land (over-grazing), fisheries (over-fishing) and the atmosphere (excessive pollution).

Such cases where public goods are either not provided or are overused are described by many text books in economics as instances of 'market failure'. The presence of 'market failure' is taken as the first step in the logic that concludes with the policy prescription that government has to step in to protect the environment.

Actions justified with recourse to the notion of market failure include the direct provision of environmental goods and services

---

1 Hardin, G. (1968), 'The Tragedy of the Commons', *Science*, **162**: 1243–1248.

through the ownership and management of environmental assets. Hence governments around the world hold portfolios of property that are declared as National Parks or nature reserves of various types. Governments in some jurisdictions are also the 'owners' of all wildlife. In these cases, rights to the assets are defined (the land, the animals) and legal ownership is achieved, but defence of those rights to exclude non-payers is problematic. Furthermore, rights to some of the goods and services that flow from the assets (such as the pleasure people enjoy when an endangered species is 'saved' from extinction) may be impossible to define.

Governments also regulate the use of privately owned assets to ensure that they are managed in ways that produce environmental public goods and services. For instance, the clearing of vegetation may be restricted on some categories of privately owned land to prevent soil erosion from impacting on people downstream and to provide 'off-reserve' habitat for species. Regulations are also used to manage the way 'un-owned' environmental assets (such as air and water resources) are used by private entities. Limits are enforced on the amount and type of pollutants that can be released into the environment.

Money-based incentives are also used by government to influence peoples' use of environmental assets. Instead of applying regulations to ensure that private owners of assets use them in ways that supply environmental goods and services, governments may pay for them in a way that simulates a market. For instance, private landowners may be invited to submit bids in a type of auction system in which they offer to provide environmental goods and services in return for payments.[2] Alternatively, farm owners may be paid for undertaking pre-defined

---

2   Examples of such 'auctions' include BushTender and EcoTender in Australia http://www.dse.vic.gov.au/conservation-and-environment/biodiversity/rural-landscapes/bushtender; http://www.dse.vic.gov.au/conservation-and-environment/biodiversity/rural-landscapes/ecotender; and the Conservation Reserve Program in the USA http://www.nrcs.usda.gov/programs/crp/.

management actions that produce environmental goods and services.[3]

Payments for environmental services (PES)[4] involve asset owners being paid for the use of their assets in the production of goods such as endangered species protection and services such as watershed protection. The implication is that the use of their assets for these purposes means that they cannot be used for the production of other goods and services (such as food, fibre and timber). The PES are therefore meant to compensate the owner for the use of their asset in ways that reduce their potential for generating private income from goods such as food, fibre and timber. For example, farmers must give up some income if they are to plant trees on their land to improve the quality of water flowing from their property into a drinking water reservoir instead of growing crops or grazing livestock. This asserts the private property rights of the landowners. In some circumstances, people may be forced by government to pay for their use of an environmental asset. Taxes paid on a per unit basis for the release of a pollutant into the atmosphere is an example. In such cases the individual is deemed not to have the right to use the atmosphere. The property rights to the atmosphere are thus conferred on the state.

Government imposed 'offset' schemes imply a similar definition and allocation of rights. These schemes operate so that any damage to the environment caused, say, in a development project is required by legislation to be 'offset' by an environmental improvement elsewhere. For instance, if a new mine is being proposed, any vegetation that is proposed for clearing may be 'offset' by the planting of trees in a nearby location that has similar ecological conditions. A 'no-net-loss'

---

3   For example, payments to farmers in the European Union are made on the basis of specified conservation actions such as the establishment of environmental 'buffer zones' around cultivated areas: http://europa.eu/pol/agr/index_en.htm.
4   See Wunder, S. (2005), 'Payments for environmental services: Some nuts and bolts' http://www.cifor.cgiar.org/publications/pdf_files/OccPapers/OP-42.pdf for an explanation of PES schemes, including their use in developing country contexts.

requirement may mean a greater area of new plantings is required to offset the loss of an established forest.⁵ Where the goal is to protect the habitat of a particular species, specifics relating to the age composition of vegetation may be important in defining an appropriate offset. Additional investments in species protection measures may also be required as a condition of the offset.⁶

The role of the private sector
As well as providing a rationale for public-sector involvement in the provision of environmental goods and services, the public good and hence market failure 'logic' has also formed the basis for the thinking that the private sector should be excluded as a potential supplier. Following on from this public good 'logic', the motivation of the private sector to make profits is argued to drive people to choose resource development options over environmental protection. Hence the conclusion is drawn that the private sector is an unreliable vehicle for encouraging conservation.

So the private sector is viewed as the cause of environmental degradation rather than having any role in the protection of nature.

However, as some of the examples of government involvement in nature protection outlined above demonstrate, there are cases when policy measures engage the private sector in the provision of environmental goods and services. Increasingly, the capacity of the private sector in supplying environmental protection, along with the incentive power that market forces can provide, are being recognised by policy makers.

---

5   In the case of Xstrata's Mount Owen coal mine in Australia, an offset of 415 ha was created for the mine's impacts on 94 hectares: http://www.xstrata.com/sustainability/casestudies/managing-biodiversity-at-mount-owen-coal-mine/
6   The endangered cassowary in far north Queensland provides an example: http://www.ellabay.com.au/pdfs_seis/D._Ella_Bay_SEIS_Volume_1/1.3_Cassowary/Cassowary.pdf

This has come in part from the recognition that the public good market failure 'logic' and the implied inability of the private sector to deliver environmental protection is not the end of the conceptual story. In addition, there is an acknowledgement that government action is not without its difficulties. These difficulties arise because of a number of technical challenges and questions about incentives.

First, there are the technicalities involved in establishing government policies. To make sure that policies will lead to an improvement in the well-being of society requires a lot of information. Knowledge of the biophysical 'cause-effect' relationships is needed to understand how alternative policy 'causes' will affect the environment. How members of society value those environmental impacts must also be understood, along with the costs that any policy measure will create. These are complexities that are often not well understood and, in addition, they are likely to be in a state of constant change.

A further element in the knowledge 'puzzle' facing governments is that people may hold information which they are unwilling to reveal for fear of making themselves worse off. Or they may want to distort their privately held information in an effort to swing policy decisions in directions that will favour them. For example, people looking to government to pay them for providing environmental services may deliberately overstate the costs they face in being 'environment suppliers'. Revealing their costs of production may also be bad for business if they are in competition with other potential suppliers. Users of environmental goods and services also have incentives not to reveal the true extent of the value they enjoy from environmental protection. By exaggerating the benefit they enjoy, they may be able to convince governments that more should be provided at taxpayer expense.

All these complexities mean that governments may not be able to gather enough timely and accurate information on which to base their policy deliberations and decision-making.

Second, there are incentive issues. Even if there was perfect information available to policy makers, would they choose options that were in the best interests of society? Experience suggests not. The temptation for policy makers (in a democracy) is to choose options that further their interests rather than the interests of the society they represent. Those interests are frequently centred on the establishment and maintenance (or even enhancement) of political power. The implication is that policies will be selected that favour select groups within society that can deliver 'swinging' votes. Such policies may come at a cost to other groups in society, but those costs will be spread so thinly across the rest of society that voters are not swayed when they are imposed. Alternatively, they may cost groups within society whose votes are already deemed to be 'lost'. Put simply, the incentives of politics are not necessarily aligned with society's best interests.

Third, there are no guarantees that the civil service will provide a brake to the political motivations of their ministerial masters. Professional public servants will have their own set of incentives. These, too, are not necessarily aligned with the best interests of society. If the senior members of the civil service are political appointees, then political interests will filter across to motivate the professional ranks. If they are not, then personal incentives for career betterment may be focused on matters such as achieving greater prominence for their own departments. Individual interest may also drive behaviour. For instance, people with a personal interest in environmental protection will be more likely to seek jobs in environmental agencies rather than departments of defence or finance. Opportunities then emerge to further their own interests through the policy advice they provide. Again, the incentives of the bureaucrat are not necessarily aligned with the best interests of society as a whole.

As a result, government actions may be ill-informed and inappropriately motivated. The implication is that even though the private sector may face problems in supplying environmental goods and services at a level that is 'optimal' for society, it may be able to do a better job than the public sector. Or there may be ways for the private sector and the public sector to combine their respective strengths and avoid their respective weaknesses to do an even better job.

In some cases this recognition of the relative merits of private sector engagement may require governments to reduce their own environmental conservation efforts in order to increase the provision of environmental goods and services overall. Existing government actions may be 'crowding out' private-sector initiatives. For example, government conservation initiatives may be sufficient to prevent private enterprises or private not-for-profits setting up to provide similar services. It may also be the case that government regulations squash private incentives to supply environmental services.

The first step in an exploration of the private provision pathway to environmental protection involves looking more closely at the 'conventional wisdom' that is at the heart of the 'market failure' analysis of public good provision.

Ownership

The use of the expression 'market failure' misses the point. It's not the market that fails per se. Markets can only 'succeed' if there is a strong foundation of property rights to available scarce resources that allow the exchange process to start. Without rights to define ownership and the subsequent defence of those rights, markets will not even form.

So it is critical in understanding why markets 'fail' that we analyse the reasons why property rights are not well-defined and/or defended in the case of environmental goods and services.

People have incentives to see property rights established. If ownership can be established, the opportunities to gain from the buying and selling of rights become available. People will therefore seek out ways to define and defend rights. The problem is that these efforts involve the use of scarce resources and are therefore costly. Unless the net gains expected by a person from the exchange of rights exceed the costs paid in setting up those rights and making the exchange, it will not be worthwhile for the person to embark on the process of establishing ownership and trading.

The recognition of these costs – known as 'transaction costs' – demonstrates that, in some circumstances, the reason why markets don't form is simply because the gains from their formation do not warrant the benefits that are generated. But this analysis considers only the costs and benefits of individual action. Can some form of 'collective' action, by which individuals group together to secure the gains available from exchange, prove worthwhile?

Collective action alternatives

Observing the structure of contemporary markets shows the importance of collective action. It is almost always the case that property rights are defined and defended through the instruments of collective action. Notably, people establish representative bodies that form organisations responsible for laws that define rights (parliaments) and other organisations (police, courts) that enforce them. The economies of scale these organisations can achieve by spreading the high costs of setting up standardised systems across huge numbers of exchanges mean that per-unit transaction costs can be dramatically reduced compared with those that would be expended by individuals acting alone.[7]

These collective action 'rules of behaviour', such as contract law and the penal code, that are created through the formation of organisations external to society itself, are known as 'external institutions'. Coupled with other institutions that come from within society, such as honesty and trust, they act to lower the transaction costs associated with exchange.

What this shows is that public-sector actions in setting up the definition and defence of property rights can be critical in determining

---

7   This illustrates that private-sector actions in markets to allocate scarce resources within an economy is actually heavily reliant on some basic collective (or public sector) actions.

whether or not a market can form and then operate successfully. If the underlying 'institutions' are set up with the specific aim of lowering the transaction costs associated with exchanging environmental goods and services, then private-sector engagement may follow. The prospect is that the public good characteristic of 'non-excludability' in environmental goods and services can be overcome through the public sector action to define and defend rights.

An example of this is the establishment of cap-and-trade schemes for pollutants. Under these schemes, governments create 'rights to pollute' in the form of tradable emission permits. Each permit allows its owner to emit a pre-defined amount of pollution. That establishes a well-defined property right. If monitoring shows that the permitted amount of pollution is exceeded, a fine (or even a custodial sentence) is imposed. That defends the right. Because the permits are valuable – they represent access to a resource (the environment) that the definition of rights has made excludable and hence scarce – trade between competing polluters will take place and allocation to the highest value user of the environment will follow.

The same logic underpins 'individually tradable quotas' (ITQ) for fish catches. There, governments set up rights to catch specific species of fish in designated areas. Each fisher must own an ITQ in order to catch and hence sell that type of fish. This property right is defended by government policing agencies when the catch is brought to shore. The ITQ are therefore valuable because they represent the right to the scarce fish resource. Market trading in ITQ ensures that the fisher who can secure the most profit from the catch (and that means the lowest cost, most efficient operator if the price of fish is the same for all) will be able to buy the most ITQ.

Other forms of collective action do not involve these sorts of high-level, government-sourced institutions that are required to facilitate formal market exchanges.

For instance, consider a group of people trying to work out who-gets-what of a scarce resource. If the group is small in number, then setting up the rules (institutions) that underpin the resource use choice process may come from within the group and at a relatively low cost. Such internally negotiated solutions are also afforded when the membership of the group is not only small but also relatively homogeneous. With small numbers in a group and with the people in the group having similar likes and dislikes, the transaction costs of negotiating an outcome and then enforcing it tend to be low enough to see agreements being struck successfully.

For these sorts of groups, the transaction costs associated with formal establishment of externally defined and enforced property rights are higher than setting up internal rules that guide resource use. The work of Nobel laureate Elinor Ostrom[8] has focused on the way these sorts of collective agreements for the provision of environmental goods and services have overcome the non-excludability characteristic: Hardin's 'tragedy of the commons' is negated by Ostrom's 'institutions for collective action'. For instance, the management of cattle on a village common can be organised to make sure that the grass is not over-grazed through the application of access rules that are agreed to by all and mutually enforced. Determining those rules and then enforcing them is so much easier (and cheaper) when the number of people wanting access to the common land is small and everyone has shared interests.

However, even with this form of negotiated collective action and larger scale collective action to define and defend property rights, the transaction costs associated with environmental goods and services may still be too high in some cases to allow exchange to be worthwhile. For instance, a protected natural area may have well-defined property rights because legislation lowers the transaction costs of establishing land title. However, commercial operation of that land to supply environmental services requires the policing of the established right

---

8   Ostrom E. (1990), *Governing the Commons: The Evolution of Institutions for Collective Action*, New York: Cambridge University Press.

so that non-payers are excluded: The right has to be defended. The transaction costs associated with such enforcement (such as fencing, patrolling and prosecuting offenders) may be so high that profitable operation is not possible. Likewise, ownership over a particular fish stock may be possible to define using ITQ, but the migratory nature of the species through international waters may preclude profitable enforcement. In other words, laws that establish rights will be ineffective where defending the rights is infeasible. No matter how 'illegal' acts of pollution, fishing or logging might be, markets will not be able to work effectively without enforcement: That is where the 'paper' of the law translates into action.

Where the transactions costs of private negotiation or market interaction are still too high to allow exchange, the more overt forms of governmental collective action, such as regulations and public provision, come into consideration. However, it is a mistake to think that these forms of public-sector involvement allow us to be free of the burden of transaction costs that weighs down private-sector engagement. The types of transaction cost may change but they exist nonetheless. Public-sector policies bear the transaction costs associated with the collection and analysis of information to design the policy, the political lobbying to have it installed, the bureaucratic processes of establishment, monitoring and policing, plus all the costs borne by those who are affected by the policy associated with compliance, evasion and enforcement.

Transaction costs represent the use of scarce resources. They are no more or less significant in choices than the costs associated with the provision of environmental goods and services themselves. Decisions made without them being considered may well lead to choices being made that leave society worse off.

Policy choices

The implication of this analysis is that there are numerous ways in which society may organise itself to see environmental goods and services produced – and then consumed. These range from private exchange in markets through collective negotiation to overt public sector action. Which of these alternative institutional arrangements (or mixtures thereof) is best suited to delivering environmental goods and services in the best interests of society as a whole? The answer depends on the specifics of the case at hand, especially the extent of the transaction costs that apply.

Recognising that public-sector provision or regulation is not necessarily guaranteed to deliver better outcomes than the alternatives of collective or private-sector action is important in establishing this policy choice process. This is especially important because the prior existence of publicly provided environmental goods and services can mean that alternative (and potentially superior) collective action or private options are prevented from emerging: They are 'crowded out'. Legislative regulations can also act to stifle technological innovations that would enable cost-effective exclusion of non-payers. In that way, private-sector provision options are not allowed to develop.

Some examples of private and collective supply options in a number of different contexts illustrate their potential.

Terrestrial ecosystem protection

Over the last few decades, there has been a considerable expansion in the area of privately owned and operated nature reserves in numerous countries around the world. The *modus operandi* of many of the organisations behind this trend is of a 'not-for-profit' business seeking donations from individuals and corporates to fund their purchases of land deemed suitable. In this way, networks of reserves outside the 'national park' estates have been created.

In Europe, groups such as the Landesbund fur Vogelschutz[9] and the Bund Naturshutz[10] organise geographically local membership branches

---

9   http://www.lbv.de/
10 http://www.bund-naturschutz.de/

to raise money to purchase properties within their region. In this way they seek to avoid the 'free-rider' problem of collective action through keeping group numbers small and homogeneous. 'Peer group pressure' for all to be engaged and contribute to the conservation cause is more readily applied in these groups, much as Ostrom predicted to be the case.

The Nature Conservancy operates at a different level. Based in the USA, The Nature Conservancy seeks to purchase property there (and also internationally) in areas where particular threats to biodiversity protection have been identified and where local sources of funds (and organisations to collect them and manage purchases) are more limited or where biodiversity values are exceptional.[11] Funds are raised using partnership arrangements with major corporates but also through extensive advertising campaigns that target individual donations.

Earth Sanctuaries Pty Ltd (ESL) was the pioneer organisation to buy protected areas in Australia. Initially its operations were funded through the sale of shares to members of the general public (particularly people who visited the company's properties).[12] As a private company that was not listed on the stock exchange, issuing new stock was unregulated and the price charged was at the discretion of the company. However when ESL sought listing on the Australian Stock Exchange as a public company, the corporate regulator (the Australian Securities and Investments Commission – ASIC) required the issuance of a prospectus that conformed to its standards. This proved challenging for ESL in providing asset values for its properties that both met accounting standards and substantiated the price of the shares being floated. After its public listing, the ESL share price gradually declined to the point where insolvency appeared likely. At that stage, the majority of its properties were purchased by another private-sector nature protection entity, Australian Wildlife Conservancy (AWC).[13]

---

11 http://adopt.nature.org/
12 http://pc.gov.au/research/commissionresearch/?a=8303
13 http://www.australianwildlife.org/

Initially funded through the benevolence of a single individual, AWC has broadened its reach to include donations from the general public. It now holds a portfolio of 20 reserves amounting to an area of 2.8 million ha.

The operations of another not-for-profit private entity, Bush Heritage Australia[14] (which owns and operates 32 reserves of almost 1 million ha) parallel those of AWC. Together they provide a significant contribution to the terrestrial nature protection effort in Australia. Their appeal is in their capacity to extend the range of ecosystem protection beyond that provided by the state. Contributors are those whose demands for nature protection are not satisfied by the public offering. This commonality of preference (a homogenous group) is used by these organisations in their marketing efforts. For instance, advertisements are taken out in magazines that attract a readership of conservation-minded people (such as *Australian Geographic*[15] and *Wild*[16]). Contributions are marketed as having a direct impact on supply, with individual fund-raising campaigns aimed at the protection of specific areas. Donors to specific projects are acknowledged to provide public recognition.

The operations of these not-for-profits use the existing institutional arrangements that underpin property markets to secure their assets. The environmental goods and services then supplied from the land assets are provided as non-excludable. Voluntary donations are thus made contrary to the free-rider incentive. The fund-raising mechanisms devised by the not-for-profits have been designed to reduce those incentives.

In addition, they seek to promote an image of being a trusted mechanism for donating through which outcomes can be assured without excess funds being diverted to administrative costs.

Some of the actions of the not-for-profits also attempt to link the

---

14 http://www.bushheritage.org.au/
15 http://www.australiangeographic.com.au/
16 http://wild.com.au/

supply of excludable private goods to environmental public good provision. In part, the publicity provided to an individual or corporation when they donate to the not-for-profit conservation organisation is a private good. This is the same logic that underpins private sponsorship of sporting teams. The sale of merchandise – t-shirts, coffee mugs and the like – to supporters also links a private good with the public good of nature protection: The supporter purchases a private good such as a 'Save the (insert your favourite endangered species here)' t-shirt and in doing so provides funds for the protection of that species. Where visiting a protected area is associated with a donation/payment for the visit, the private good provided by the visit is produced alongside the non-excludable benefits of nature protection such as endangered species survival. Just as the production of meat and leather comes from growing cattle, the visit and the species protection are provided as 'joint goods'.

With public sector assistance
The activities of many not-for-profit environmental entities take advantage of the institutions established and funded by the public sector such as the land title office. Security over the title to land purchased for conservation is thus ensured. In some cases default ownership (in the case of liquidation of the not-for-profit asset owner) may also be assigned to the state as a reassurance to donors that even in the event of a collapse, the paid-for conservation values will not be lost to a development project. This assignment may be in the form of a covenant over the title where the beneficiary of the covenant is a state agency.

However, not-for-profits are often also advantaged by other public-sector initiatives. In these cases, the private and public sectors cooperate to supply environmental goods and services. The public sector may be a financial contributor to the private-sector activities. In doing so the state acts to mobilise funds from the general public (through general taxation) who may be argued to 'free ride' on the benevolence of those who do make private donations.

For instance, the acquisition by AWC of the ESL portfolio of conservation properties was funded on a dollar-for-dollar basis by the Australian Government: For every dollar of privately-raised AWC funds, the state contributed a dollar.

Tax deductibility of donations is also sought by private-sector entities. This effectively lowers the 'after-tax' cost of a donation made by a tax-paying individual or corporate and thus is an effective stimulant to donating behaviour. Thus tax-deductible status is valuable for not-for-profits and not having it is a considerable barrier to start-up operations.

Riparian ecosystem protection
Although the operation of not-for-profits in the provision of terrestrial ecosystem protection is well established, the same is not true where water rights are involved. Partly this was because the external institutional structures necessary to facilitate markets in water rights had not formed privately and were not (at least in many jurisdictions) established by the state.

The importance of well-defined rights to water in the protection of riparian ecosystems is illustrated by environmental protection initiatives taken in the Murray Darling Basin (MDB) in Australia.

A key plank of the water reform process undertaken in the MDB over the past decade was the separation of entitlements to extract water for irrigated agriculture from land title. Once separate title to water was established, water trading became prevalent. This enabled water entitlements to be transferred from lower value uses to higher value uses within agriculture. Water formerly used to irrigate wheat crops, for example, was bought by downstream farmers wanting to grow grapes.

However, trade in water entitlements was not limited to farmers. Conservationists also recognised the possibilities offered by the water market. Not-for-profits such as Healthy Rivers Australia[17] were created as

---

17 http://www.healthyrivers.org.au/

vehicles to accept donations from members of the public and corporates that could be used to buy water entitlements that could then be used to water wetlands and floodplains instead of crops. Donations from irrigators of their seasonal entitlements to water were also accepted by Healthy Rivers Australia. Some of these irrigators were no doubt keen to show their 'corporate social responsibility' while others may have simply been seeking the enjoyment of the environmental benefits that ensued.

Australia's peak environmental lobby group, the Australian Conservation Foundation, also entered the water market to secure benefits from a specific riparian site.[18]

The opportunity to use the water market to re-allocate water from irrigated agriculture to 'environmental flows' has also been taken by governments. Declining river health – largely caused by the over-allocation of water to irrigation – caused governments to seek ways of reallocating water that would see farmers being compensated for their lost assets. 'Buy-backs' in the water market were the vehicle chosen. The federal government established the Commonwealth Environmental Water Holder[19] (CEWH) to enter the water market to purchase entitlements and then be responsible for the management of the asset it held to achieve environmental goals.

The action of the state in entering the water market is also a good illustration of the 'crowding out' effect on private-sector initiatives. The purchases of entitlements made by the CEWH were so extensive that the extent of unsatisfied demand for environmental water amongst the general public was greatly diminished. Hence the prospects for private not-for-profits to secure donations, despite their tax deductibility, were much reduced.

The establishment of the CEWH also illustrates one of the hazards of public-sector provision. If the actions of CEWH effectively crowd

---

18 http://www.acfonline.org.au/articles/news.asp?news_id=2744
19 http://www.environment.gov.au/water/policy-programs/cewh/index.html

out any private sector 'competitors', the danger is that it then acts as a monopolist. This would mean a bloated cost structure and an unwillingness to operate in ways that respond to the people's demands for environmental goods and services. Without the discipline afforded by competition, the bureaucrats in charge would only have the incentive to further their personal interests.

Further prospects
There remain many other prospects for private-sector engagement in the provision of environmental goods and services in both developed and developing countries.

Some of these opportunities involve the joint production of public and private goods. The Property and Environment Research Centre (PERC), through its 'enviropreneurs' programme, specialises in identifying these and developing them.

For instance, hunting and fishing are activities that can be organised in ways to exclude non-payers. And the hunters and fishers who do pay and have access to the resource then have the incentive to protect that asset along with the environment that supports it. Environmental protection is afforded through the hunting and fishing activities. Big game hunting in some African countries has been the conduit for a number of species conservation initiatives. European and North American hunters are willing to pay enough money for their sport to provide an income source sufficient for the local people to want to protect their game animals and their habitat.[20]

One danger to the emergence of such entrepreneurial activities is the regulatory action of government. For instance, the Convention on International Trade in Endangered Species of Wild Fauna and Flora (CITES)[21] aims to protect endangered species by restricting their sales. Ironically, this effectively limits prospects for the commercialisation of

---

20 http://www.perc.org/articles/article316.php
21 http://www.cites.org/

some species, and their products, that would deliver strong incentives for their protection. For instance, trade in rhino horn is currently restricted to the black market and supplied primarily by illegal poachers who kill the rhinos to remove their horns. Commercial rhino horn 'farms' that involve the sustainable harvesting of the rhino horn without harm to the animal are thwarted by CITES.[22]

Further examples of 'enviropreneurial' initiatives range from the provision of burial sites for people in natural protected areas[23] through to using the fur of a pest animal introduced into New Zealand, the possum, in the production of garments and at the same time giving an incentive for their numbers to be controlled.[24]

Recreational use of natural areas also offers opportunities for the exclusion of those who don't pay for the supply of species protection. Low-cost methods of exclusion are possible even in remote areas where boundaries are difficult to enforce. For instance, pre-purchase of entry permits coupled with random checks by mobile patrols can increase the probability of non-payers being caught. This in turn increases the proportion of people who will choose to pay for their use.

Commercial, privately run facilities and services associated with the use of protected areas can also provide a source of funding for the management of such areas. Hence, operators of hotels, camping grounds, kiosks, restaurants and guided tours provide 'joint goods' alongside the public environmental goods and services. The two forms of production can be linked so that the commercial incentives 'spill over' to the supply of public goods.

In a similar fashion, commercially profitable recreation facilities near urban areas may be used by a private supplier to cross-subsidise the provision of more remote but more environmentally valuable sites.

---

22 http://www.perc.org/articles/article1409.php
23 http://www.perc.org/articles/article1015.php
24 http://www.perc.org/articles/article1066.php

Allowing the private sector to own and operate the portfolio of national parks – the 'privatisation model' – may be a difficult concept to introduce politically. An alternative would be to introduce public/private partnerships into the environmental protection sphere. This would be an extension of the idea common in the areas of transportation and energy supply. Although the state owns the asset, the private sector operates it. This merges with the idea of the private sector operating facilities within protected areas.

Key to the success of private-sector interests managing state-owned protected areas is the injection of competitive pressure in the management operations. This would require periodic tendering for the management task in a competitive setting.

All of these options require that governments give up some of the roles that have commonly been deemed to be their 'natural' assignments. Only if governments step aside will private-sector action become financially feasible.

Similarly negotiated solutions to environmental choices may be more forthcoming if governments stand aside from the role of regulatory authority. If governments do not determine an outcome, individuals will find themselves facing the prospect of developing an outcome by negotiation. For instance, in the context of a water pollution problem, rights may be defined so that the polluter must pay for damage caused or the polluted party must pay to have the pollution reduced. If there is no prior legal standing on the issue, small groups of affected parties may negotiate an outcome. It may be that the polluted party finds it a lower cost option (including transaction costs) to pay the polluters to stop their activities. A government ruling that enforces the 'polluter pays' principle would, in that case, mean the least-cost option is forgone. In such small group cases it may be better for government to let private negotiation sort out the resource use choice.

A 'little green lie'?

Environmental policy in most developed countries is heavily influenced by the logic of 'market failure'. The result has been a heavy dominance of government in the supply of environmental goods and services, both as a direct provider and as a regulator.

However, many examples demonstrate how the private sector can be involved. Furthermore, the consideration of transaction costs and incentives shows that, in some contexts, it is preferable for the private sector to take a larger role. This would indicate that the calls for exclusion of the private sector in matters environmental are supported by a 'little green lie'.

Taking advantage of the strengths of market forces is a key part of engaging the private sector in environmental protection. But the so-called 'market-based instruments' such as taxes, subsidies and permit trading schemes that have become increasingly popular in environmental policy circles are only a small part of the story. Rethinking environmental goods and service supply to recognise the potential of negotiations between parties in small groups could see governments backing away to allow lower cost options to be explored. The joint production of private and public goods associated with the environment also has the potential to see greater private sector engagement. These types of arrangements already exist and they work. They could be expanded if government allowed them the 'room' to develop.

With these advantages available, the motivations for the 'little green lie' associated with exclusive public sector control are not readily identified. Certainly, those in the public sector – both politicians and bureaucrats – have incentives to see their roles perpetuated and enhanced. Those with an ideological conviction against the private sector will also take the opportunity to argue in favour of absolute state control. The irony of this position is that by allowing greater private sector involvement, more environmental goods and services may be produced. This has clearly been the experience in not-for-profit provision of protected terrestrial ecosystems.

One objection to private sector involvement is that the owner of an environmental asset may charge an access fee whereas the government may provide it free of charge. Frequent users may use the public sector 'little green lie' to further their lobbying for state supply in an attempt to maintain free access. It may also be argued on equity grounds that poorer people should not be excluded from access because they cannot afford to pay the price. This argument is increasingly flawed by governments finding it necessary to charge access fees in order to cover costs. It is also exposed in situations where no price is charged and congestion results: users then bear costs (particularly in terms of time spent waiting in queues) other than an access fee. And the issue of equity is also best addressed using measures that are not based on price – for instance, the provision of a quota of 'free access passes' on a means-tested basis.

11: Agriculture and Mining

Proposition: Agriculture and mining are always in conflict with the environment.

Agriculture and mining are extractive industries which deplete our stock of natural resources, often irreversibly. They also cause environmental degradation including soil erosion, biodiversity loss and contamination of the gene pool, water and air.

BUT

While there are some trade-offs between agriculture, mining and the environment, these can be reduced through the use of management techniques and technologies. Offsets and remediation work on farms and mines can maintain and even improve the condition of the environment.

Pros and cons

There is no doubt that agriculture and mining change the environment. To grow crops requires existing vegetation to be cleared, the soil to be cultivated and potentially fertilised. Exotic species are necessarily introduced into the ecosystem. Some of these are unwanted pests that require control. The water cycle may be altered. To graze livestock means that the mix of plant species present will change, especially if new pasture species are introduced. The soil and water regimes will also be changed. Underground mining may have impacts on aquifers and the development of product and waste stockpiles that affect surface ecosystems.

Open-cut mines have larger surface 'footprints'.

Not only do mining and agriculture affect the environment at their immediate sites but they can also cause changes 'off-site'. For example, extractions from rivers for irrigated crops have down-stream effects and the dust from mining operations can blow on to neighbouring areas.

Generally, these environmental impacts are considered to be detrimental. Yet as a society we continue to choose to engage in mining and agriculture simply because we value the goods and services that flow from those activities. Food, clothing, energy and the structures in which we live and work are core amongst the products that flow from the labours of miners and farmers. Indeed it is difficult to imagine how humanity could have developed beyond the very rudimentary or how it could continue even to survive without these 'primary' industries.

However, the 'can't do without it' argument applied to mining and agriculture is reminiscent of the 'infinite value' of the environment 'little green lie' rehearsed in Chapter 8. The issue is not if we can survive without either the environment or the products of activities such as mining and agriculture. Rather, it is a matter of society deciding where to establish the trade-off between the pros and cons of mining and agriculture: Should we have a bit more farming, knowing that to do so would mean a bit less environmental protection? Or, should we not go ahead with a new mine in order to protect some more of the environment?

That decision is easy if the population is starving and cold because so little of our resources are devoted to farming and mining because they are mostly set aside as protected environmental areas. With so much protected environment, losing some of it to mining and agricultural development would not be of much concern relative to the value that society would enjoy from being able to gain some respite from their survival woes. The trade-off decision is equally straightforward if environmental conditions are severely degraded because of mining and agricultural activities: a shift toward more environmental protection is likely to increase the well-being of society as a whole.

Agriculture and Mining

In both cases, more agriculture and mining means less environment and *vice versa*. Hence a 'win' for agriculture is a 'loss' for the environment, and a 'loss' for mining is a 'win' for the environment. This picture of a trade-off between development and protection is, however, overly simplistic. Importantly, it may be possible to envisage circumstances in which both agricultural and mining production and environmental conditions improve. Rather than a trade-off, this would be a 'win–win' situation.

'Win–wins'

The first context in which such trade-offs or 'win–loss' contexts can be swapped for 'win–win' conditions occurs in the agricultural sector when environmental conditions are already severely degraded because of past management actions.

Agriculture, at a fundamental level, relies on the environment as an input. Below some threshold level of environmental quality, agricultural productivity declines along with the condition of the environment. For instance, water quality may have been so reduced through the intense application of chemical fertilisers, pesticides and herbicides to the point where it adversely affects crop or livestock productivity.

The implication of this interrelationship is that the opposite applies in this context: removing some of the pressure on the environment due to agricultural activities may improve both agricultural production and the condition of the environment.

For instance, lowering the intensity of livestock grazing in circumstances where over-grazing has been the norm may increase pasture cover so that agricultural productivity increases and water and air quality are improved because of lower soil erosion rates.

This type of 'win–win' relationship will hold only if environmental 'tipping points' have not been exceeded. For instance, if the soil has been eroded by over-grazing to the point where pasture recovery is impossible, then the removal of the grazing pressure will not allow the recovery of agricultural productivity or environmental health. In such circumstances, agriculture ceases to be sustainable into the future. Its use of resources becomes an irreversible activity and so the resources involved become non-renewable. In the same way, resources exploited by mining have rates of formation that are a very small fraction of the rate of extraction and are therefore non-renewable.[1]

The 'win–win' prospects afforded by reducing resource-use pressure in agriculture when resources are already over-used and so degraded can be enjoyed only up to a point. Decreasing the intensity of agricultural production will keep on improving environmental conditions but, eventually, it will also lead to decreased overall agricultural output and the familiar 'trade-off' issue re-emerges.

However, even then there is potential for another source of 'win–win' outcomes to emerge. This applies to both mining and agriculture and occurs when more inputs become available for use in the production process. For instance, if more land becomes available for agriculture, pressures on existing farming land may be reduced sufficiently to see environmental conditions and overall farm production increase.

Perhaps of more relevance, however, is the prospect of improved knowledge and technology being introduced into agricultural and mining operations. This involves more 'human capital' being used in the production process. For instance, the discovery and implementation of new mining or mineral processing techniques can increase productivity and do so while improving environmental outcomes, or at least without doing further environmental harm. Other innovations may allow improved environmental outcomes without any reduction in mine output. Improved mine site management – including noise abatement, dust suppression and waste water treatment – fall into this latter category.

---

1 Because mining activities are not so directly dependent on environmental conditions for their productivity, the prospects are weak for improving mine output along with better environmental quality (the 'win–win' opportunity) even when the environment has been severely degraded.

In agriculture, the close interaction between production and the environment has ensured that knowledge and technology have important roles in developing 'win–win' outcomes. An example is the development of zero (or minimal) tillage cropping. Common past practice in grain growing involved burning the 'stubble' remaining after the harvest of the previous crop followed by intensive ploughing to prepare the soil for the next crop. Increasingly, stubble is now retained, the ground is not ploughed and the seed for the next crop is directly 'drilled' into the soil. This ensures that soil disturbance is minimised. The impacts of these changes have been to increase crop yields because of improved soil structure and nutrient retention as well as to lower rates of soil erosion and resultant downstream environmental problems. Reduced cultivation also results in lower fossil fuel consumption and associated emissions of carbon dioxide.

Much of the success of zero tillage farming has been due to the development of environmentally benign herbicides, such as the glyphosates, and genetically modified (GM) crops that are not affected by those herbicides. Together, these technologies overcome the problem of weeds in crops. Without having to plough to control weeds, farmers have maintained or improved yields while reducing the costs associated with cultivation. This has proven to be a profitable strategy for developing and developed country farmers. Hence, the practice has been widely adopted. The result has been a true 'win–win'.

Other agricultural chemicals developed to improve yields have not been so kind to the environment. An example is the use of insecticides based on the organochlorine chemical endosulphan to control infestations of the boll weevil in cotton crops. Widespread (even aerial) application of these insecticides was successful in preventing yield reductions due to infestations, but the associated contamination of water sources resulted in human and environmental health problems. Endosulphans bio-accumulate up the food chain and can cause reproductive and developmental damage in animals.

In the case of the use of endosulphans, the technological development resulted in a trade-off between agricultural production and environmental condition.

Such was the environmental risk posed by endosulphans that they have now been widely banned. Fortunately, an alternative insect-control technology was developed that avoided the agricultural/environmental trade-off. This involved the development of GM breeds of cotton that were resistant to boll weevil attack. Planting *Bacillus thuringiensis* (Bt) cotton allows farmers to avoid the use of insecticides without loss of productivity.

Innovations in crop breeding, collectively known as the Green Revolution, increased agricultural productivity so that it was possible to produce more food from less land area. The so-called Borlaug hypothesis – named after Nobel Prize winning agronomist Norman Borlaug – is that the Green revolution was also a win–win. This is because the increases in agricultural productivity reduced pressures to clear more forest for crops.[2]

There are many other examples of agricultural innovations that improve farm production and the condition of the environment both on and off the farm. They include changes in management. For instance, when Australia was first settled by Europeans, the farming systems applied were primarily those used in the UK even though the environmental conditions were so dramatically different. The result was a steady decline in the condition of the environment. Soil structure was lost and both wind and water soil erosion became problematic. River health deteriorated as too much water was extracted for irrigation. However, with the passage of time and the experience of living and working in the Australian environment, farmers have become aware

---

2   The hypothesis remains contentious given that increased crop yields and associated increases in crop profitability can also be an incentive for farmers to increase their plantings (See: http://www.sciencecouncil.cgiar.org/fileadmin/templates/ispc/documents/Publications/EIAStudy2011James.pdf)

that by adapting more to the local conditions, the productivity of their operations would increase and at the same time their operating environment would also improve. The symbiotic relationship between agricultural activities and the environment become better recognised and utilised. Stocking rates were adjusted. Instead of clearing all trees, shelter belts were maintained or re-planted. The geographic areas suited to various crops were better defined.

In some cases, communal, cooperative activity has been found to enhance farming and environmental outcomes. For instance, the Landcare[3] movement in Australia grew out of a coalition of the peak agricultural lobby group (the National Farmers Federation) and the peak environmental lobby group (the Australian Conservation Foundation) with the goal of improving the condition of communal assets that were important not only to farmers and local communities but to those who value the environment. Local area Landcare groups were formed to bring members of farming/rural communities together to work on projects such as the restoration of shared river banks or the eradication of invasive weed infestations over multiple properties.

It is important to note that the definition of what constitutes a 'win' in these circumstances depends on the definition of the starting point. For instance, the development of a new technique for mine site rehabilitation would create a 'win' for the environment relative to the no-rehabilitation option or even a previously used technique. Likewise, a new technology for the treatment of waste water from on-site minerals processing plants may provide an environmental improvement relative to the performance of the previously used technique. If comparison is made to the situation existing without any agriculture or mining, then an environmental 'loss' will always be recorded. However, such a comparison makes little sense when it is recognised that what is relevant are decisions being made now regarding farming and mining and their

---

3   http://www.landcareonline.com.au/

interaction with the environment. Current decisions should be based on current circumstances and whether or not any change from the current situation is worthwhile. What exists in all areas used for agriculture or mining is not a pristine environment. It is an altered environment. That should be the starting point for decision-making about the future.

It is not that society wants a pristine environment everywhere. Society wants some environmental protection, including some areas in which mining and agriculture does not occur at all. However, society also wants to be able to feed, shelter and clothe itself, amongst a range of other wants that require the efforts of farmers and miners. It would also appear that society has also come to want farming and mining for their impacts on the landscape and the environment. For instance, the European Union subsidises its farmers to maintain specific farming-based landscapes that support specific species of birds and insects. Land trusts in the United States have been established to protect agricultural landscapes threatened by more intensive development.[4] Even mine sites have become popular tourist destinations which local communities seek to have protected in their operating state even after mining has ceased.[5]

So the choice about the trade-off between engaging in farming and mining on the one hand and protecting the environment on the other is far from straightforward. We like to protect the environment for what it offers us but we also like to consume the goods and services provided from the development of the environment provided by farmers and miners. We like to visit protected natural areas but we also like to travel through 'traditional' rural countryside and see historical mining operations.

---

4   For example, the Montana Land Reliance (http://www.mtlandreliance.org/)
5   Queenstown in Tasmania provides an example (http://www.queenstowntasmania.com/index.php). Copper mining operations in the past denuded the hillsides surrounding the town of vegetation. This landscape became a feature of the tourist experience to the area. Similarly, both working and closed mines around the world offer visitor tours. For example, gold mines in Colorado: http://www.goldminetours.com/goldminetours.com/Home.html.

The debate therefore becomes one of where to 'draw the line' in terms of the trade-offs between protection and development of the environment. It also rests on how society can best encourage the development of 'win–win' outcomes.

Driving forces

To understand the trade-off decision and the way in which it is made, the incentives faced by farmers and miners need to be understood. It is the combination of decisions made by millions if not billions of individuals, corporations and governments that determines the overall mix of food/mining production and the environment. In turn, those incentives are many and varied but a few stand out as being instrumental.

Fundamentally, farmers and miners decide to embark on their production activities if they see that it will make them better off. If the benefits of an enterprise exceed the costs, the incentive will be to proceed. This in itself is a trade-off, as it sets up a comparison between the values enjoyed from the agricultural and mining goods and services generated against the value of the resources used. If this trade-off made by the individual involves benefits and costs that reflect society-wide values, then the choices made by farmers and miners will lead to outcomes that make the whole of society as well-off as the available resources allow. Where the property rights to the resources used in agriculture and mining and the goods and services created are well-defined and sufficiently defended to allow competitive markets to form for those resources and products, signals regarding social values are sent to the individual decision-makers via prices. The benefits of agricultural and mining goods and services are transmitted from consumers through to the farmers and miners. The costs associated with resources used – the value of the resources if they were used in their next best use – are also signalled back to the farmers and miners through the prices they have to pay for those resources.

Markets thus transmit incentives to individual farmers and miners that drive them to make socially desirable choices ... so long as the property right institutions are in place.

The costs of setting up well-defined and defended property rights to some resources used in agriculture and mining can cause divergences to emerge between what is in society's best interests and the choices made by individual farmers and miners. Because the products of agriculture and mining are primarily held under private property rights, there are established incentives for their production by individual private miners and farmers. However, the difficulties of establishing ownership over some environmental protection benefits means that there is a dilution of the incentive for market-based production. Society's desire for environmental protection may not be adequately transmitted through markets for the resources that can also be used for mining and agriculture.

The outcome is that, potentially, too many resources such as land and water are used for agriculture and mining and too few are used for environmental protection. Put simply, the trade-off is likely to be made in favour of the commercially orientated farming and mining activities.

With market forces delivering trade-off outcomes that appear to be the wrong ones from a broader societal perspective, the inclination of government has been to intervene more directly. Introducing policies that establish environmental protection areas that exclude mining and agriculture and/or regulations that limit farming and mining management practices to limit environmental damage are examples.

As a result of these interventions, the agricultural and mining sectors tend to establish lobbying power in an attempt to direct government actions in their favour, just as environmental interests do. It is the government in these circumstances that can deliver resource-use opportunities to those interested in the trade-off outcome. Influencing government through the political process thus becomes a priority. Environmental lobby groups will seek to push for 'greener' outcomes and mining and agricultural lobbyists will seek to push back on those efforts in order to maintain their access to resources and the profits that provides.

Whether or not government action delivers trade-off outcomes that are superior to those which come about without intervention remains a moot point. Certainly political power is distributed differently to the money-voting power reflected in markets. And there is evidence that the political process can end up producing more environmental harm than good.

Governments gone bad

Internationally, governments have a history of providing assistance to agriculture which in the long run has been detrimental to the environment. Assistance has been of two different types. First, there has been the provision of free (or at least low-cost) access to resources. Prominent in this regard has been state-sanctioned or direct supply of water resources. In order to encourage agricultural development, governments have allowed subsidised access to water supplied from state-funded dams and diversions, usually with encouragement from politically active farmer interest groups. The result has been excessive extractions of water (including from aquifers) to the point where surface water-based ecologies have collapsed.

The Aral Sea provides an illustration of the consequences of the over-allocation of water resources for irrigated agriculture. The Soviet government-led expansion of irrigated cotton cropping, fed by waters from the Amu Darya and Syr Darya Rivers, resulted in a dramatic lowering of the Sea's level. This resulted in raised salinity levels in what remained of the Sea with consequential negative impacts on the region's ecology and its dependent population.[6]

---

6   Glantz, M. (2007), 'Aral Sea Basin: A Sea Dies, a Sea Also Rises', *Ambio.* **36** (4): 323.

The history of Australian irrigated agriculture also demonstrates the potential for political processes to deliver an over-allocation of water to agriculture. To maintain political power, successive state and federal governments pandered to the interests of the rural lobby by building dams and diversion infrastructure and providing access to water at a price well below the costs of supply. In the state of NSW, the situation eventuated that more water was allocated as licences to farmers than existed in the rivers. Insufficient in-stream water remained to ensure the systems' long-term ecological functioning.[7]

Land degradation in western provinces of China is also illustrative of how government agricultural policy can have negative environmental consequences. During the Mao era, inland development was encouraged with state industries being established in the western provinces and populations moved from the coast. Agricultural self-sufficiency was forced on to provincial governments. The consequence was more farm output being required of areas which, from an ecological perspective, were not well-suited to agriculture. The result was over-cropping and over-grazing with resultant high rates of soil erosion. Dust storms became more frequent. The silt load in the Yellow River increased so much that, in downstream reaches, the level of the river bed rose above the surrounding floodplain. This meant that when the river flooded, the flood waters remained on the flood plain for longer periods, increasing the damage caused.[8]

In each of these cases, the consequences of intervention were not only detrimental to the environment but they also involved negative impacts on agriculture because of the interaction between the condition of the environment and agricultural production.

---

[7] Bennett, J. (2010), 'Australia's Water Reform Effort: Progress and prospects', in D. Leal (ed.), *The Political Economy of Natural Resource Use,* The World Bank, Washington DC.

[8] Bennett, J. and A. Kontoleon (2010), *Property Rights and Land Degradation in China*: http://www.crawford.anu.edu.au/pdf/staff/jeff_bennett/china_land_use/property_ rights_and_land_degradation_in_china.pdf. Viewed 17 August 2011.

The second form of government assistance to agriculture that has caused environmental concern is the protection of domestic farmers from international competition. Again at the behest of the agricultural lobby, governments have introduced policies that have sought to isolate farmers from overseas supplies. This has taken many forms, ranging from outright bans on imports, to taxes on imported products (tariffs) and subsidies paid on import-competing domestic production. For example, the Farm Bill in the United States channels funds to wheat farmers who otherwise would be uncompetitive in international markets through the 'export enhancement plan'.[9] In the European Union, the Common Agricultural Policy (CAP) involves the 'single payment scheme', under which direct payments of income support are made to farmers irrespective of their farm output. In 2010, these payments amounted to over 36 billion Euros.[10]

Such trade interventions distort the price signals provided by markets so that resources end up being misallocated. Too many scarce resources end up being used in agriculture in the countries protected from competition and too few are allocated to farming in unprotected nations. The problem with this outcome from an environmental perspective is that agricultural resources are over-used in the trade-protected areas. This over-use becomes manifest in terms of, for example, aquifers polluted with nitrates from intensive livestock husbandry and soils that lose their structure from excessive cultivation so that they become more readily erodible. In contrast, the agricultural resources of areas that are not protected are underutilised in that they could be used more intensively without causing environmental impacts.

Agricultural protectionism also has implications for the distribution of incomes internationally and this, in turn, has implications for the environment. Predominantly, protectionist policies are applied in developed countries, particularly in the US and Europe. Developing

---

9   http://www.fas.usda.gov/info/factsheets/EEP.pdf
10 http://ec.europa.eu/agriculture/markets/sfp/index_en.htm

country farmers who are unable to access these protected markets are made worse off because of the policies. With lower incomes, they are less interested in protecting their environment and less able to afford to do so. This 'income effect' exaggerates the distortions to agricultural resource allocation, including environmental resources, caused by protectionist policies.

Government policies have also been designed to assist the mining industry. They too involve lowering the costs of resource access and international protection.

Subsidies being paid to fossil fuel miners are the focus of environmental lobbyists' attention. This is because they are argued to increase the amounts of greenhouse gas emissions being emitted into the atmosphere with consequences for climate change.[11] The subsidies include infrastructure provision and fuel rebates. In the US, the 1872 Mining Law does not require the payment of any royalties for hard rock minerals extracted from public lands. The cost of this and other subsidies is calculated to be in the order of USD160m per annum. Encouraging mining on public land through this subsidisation is seen by environmental advocates as being in conflict with the use of these lands for wilderness recreation.[12]

Protectionist policies have also been extensively applied in the mining sector. As well as import-restricting measures, export restrictions may also be aimed at protecting the interests of domestic suppliers. For instance, China's decision to restrict the export of rare earths may be seen as a mechanism for manipulating the international price of these

---

11 The Green Party in Australia argues for the removal of subsidies to the coal industry that they calculate to be equal to around AUD1b per annum for the state of NSW: http://nonewcoal.greens.org.au/coal/speeches. In the European Union, it has been announced that subsidies for the coal mining industry will be phased out in 2018 instead of 2014, so giving rise to criticism from the lobby group, the European Environment Bureau (EEB): http://www.eeb.org/index.cfm/news-events/news/eu-extends-coal-mining-subsidies-end-date/.

12 http://www.pewtrusts.org/uploadedFiles/wwwpewtrustsorg/Reports/Wilderness_protection/cost_of_inaction.pdf

minerals. This is in contrast to the policy justification advanced by the Chinese Government that the restrictions are a means of containing the environmental damage caused by mining and processing and a way of holding back the rate of extraction that has accelerated with growing demands.[13]

Ironically, protectionist policies like the Chinese rare earths export bans are often advanced with environmental protection as a key justification. Preventing imports from countries where environmental protection policies are 'inadequate' can be used as a means of extending the regulatory framework of the importing country to encompass the exporting country. The 'shrimp-turtle case'[14] before the World Trade Organisation's (WTO) Appellate Body in 1998 was an example of this. Malaysia and other shrimp-producing countries appealed to the WTO for relief against the United States policy of banning shrimp imports that were caught using nets that did not include a turtle excluding device (TED). Without the TED in place, endangered species of turtles are caught and killed in the nets during shrimp harvesting. The US policy was the result of environmental lobby groups pressuring the government to apply the US Endangered Species Act on imported as well as domestic products. The WTO decision to allow the US to require a process to be used (the inclusion of a TED) provides the opportunity for environmental grounds to be used as a basis for protectionist policies.

The result has been a growth in the number and strength of coalitions forming between protectionist interest groups and environmental lobbyists. Meetings of the WTO aimed at reducing barriers to international trade (the Doha Round)[15] have been characterised by protests from labour unionists seeking to protect their employment against international competition and environmentalists seeking to use trade barriers to achieve environmental goals.

---

13 www.wto.org/english/news_e/news11_e/g20_wto_report_may11_e.doc
14 http://www.wto.org/english/tratop_e/envir_e/edis08_e.htm
15 http://www.wto.org/english/tratop_e/dda_e/dda_e.htm

Getting it right ... or at least better

Government intervention has given no guarantee of getting the trade-off right between mining and agricultural development and protection of the environment. The pressures of political 'rent seeking', including those exerted by the coalitions between protectionist interests and environmentalists, leave society with such mixed signals as to where resources are best allocated. Market signals about relative resource scarcity have been considerably distorted by government policies, particularly in the case of agriculture. This is largely due to the political significance of the rural lobby groups in the developed countries, notably the European Union and the United States: Instituting reforms that involve the winding back of agricultural protection threaten the electoral viability of any political party.

Notwithstanding these political perils, moving forward to implement the WTO reform agenda would constitute a key step in securing an improved position in the trade-off between agriculture/mining and the environment. Recognition that restricting trade is counter-productive to the achievement of environmental improvement goals – as rehearsed in Chapter 5 – would be an important start. This would also involve an acknowledgement that environmental management in many developing countries is problematic. But restricting trade is a very blunt instrument for securing its improvement. The shrimp-turtle case is instructive in this regard. The failure of fishers in developing countries to use nets with TEDs did not arise because of any inherent desire to kill turtles but rather because they were more expensive. International aid to provide nets with TEDs would have solved the environmental problem without causing the distortions to price signals resulting from import bans. Such 'environmental aid' may comprise capital equipment – like nets with TEDs – but it may also involve the provision of training, including of policy makers with the responsibility of establishing and maintaining environmental 'institutions'.

Part of that training would need to stress the significance of well-defined and defended property rights to resources used by agriculture and mining in the delivery of improved trade-offs. The evidence is clear that too many resources will be allocated to either agriculture or mining (and necessarily too few to the environment) when access to those resources is free because no-one owns them. The over-grazing of pasture land in western China is a clear example. Salinity in the soil caused by rising water tables when excessive 'free' irrigation water is applied is another. Rates of extraction of minerals that are 'free' to access are likely to be 'too fast', causing current gluts and likely future shortages.

Ensuring secure tenure to land and water allows the development of markets for those resources that better reflect their true social value. The separation of water title from land title in Australian irrigation districts enabled water markets to form. Farmers competing with each other for the available titles ensured a price that reflects the resource's next best use and so sends a signal to potential users that better reflects its relative scarcity. The productivity of water use increased as a result.[16] Environmental interests such as not-for-profit water trusts also have the potential to enter the water market to secure title so that it isn't all allocated to agriculture as outlined in Chapter 10.

Secure title to land has also demonstrated the power of delivering 'win–win' outcomes for agriculture and the environment. The *doi moi*

---

16 Security of title includes freedom from sovereign risk whereby government seizes rights without compensation being paid to their owner. For instance, where government 'takes back' entitlements to water from irrigators in order to provide environmental flows, efficient long term allocations of water are unlikely to emerge from market trading. At best, short term leases of entitlements would be negotiated: Users would be unlikely to want to buy entitlements that are subject to confiscation except at prices that would be unappealing to sellers. Insecurity over land title can cause similar problems. For instance, where legislation to ban tree clearing on freehold land has been foreseen, a race to clear as much country as possible before the regulation is enacted is started. Once in place, the regulation then becomes an egregious source of potential wealth loss for those who didn't clear and a source of resource wasting legal challenges and monitoring of illegal activity.

(renovation) reforms of land ownership in Vietnam – providing private use rights to previously communally held and operated lands – not only stimulated increases in rice production but also gave farmers the incentive to invest in the long-term productive capacity of their land. This meant changes to land management such as intercropping of pastures and crops, the planting and retention of vegetation along streamlines and the inception of zero-till methods. These investments in change reduced soil erosion, limited the progression of acid sulphate soils and mitigated soil salinity.[17]

The development of property rights to resources affected by mining would provide opportunities for markets to drive incentives for better environmental outcomes. If clear rights could be assigned to the airspace around a mine that is affected by noise and dust, negotiations between the mine owner and surrounding residents would generate price signals to which both parties could respond. For instance, if the air space was owned by the adjacent residents, the mine would need to buy access rights if it wished to continue to operate with noise and dust. The price paid by the mine for those rights would give it an incentive to reduce its noise and dust outputs. Alternatively, if the mine owned the air space rights, nearby residents would have to pay the mine to reduce its noise and dust outputs.

In some circumstances the trade-off to be developed is not simply between mining and the environment but extends to involve three dimensions: Mining, environment and agriculture. This occurs when mining affects resources that support both agriculture and the environment.

Large-scale open-cut mining presents such a circumstance when it is carried out in areas which are otherwise allocated to agricultural and environmental uses. Then, the mining operation takes agricultural land out of production. It may also affect the water table or surface water flows on which agricultural enterprises in the surrounding region

---

17 http://www.fao.org/docrep/009/ag089e/AG089E08.htm

depend. The impacts on agriculture may be irreversible given technical limits to mine site rehabilitation. In addition, the mining operation may cause impacts on the regional environment.

Property rights to the surface land (privately owned) and the underlying minerals (property of the state) land in Australia are well-defined and defended. Yet for this type of open-cut mine, the trade-off between agriculture and mining that is established by the trading of rights remains controversial. In agriculturally productive areas of NSW and Queensland where there are also rich coal deposits, farming interests have resorted to the political process to hold back coal-mining expansions. In Queensland, 'Strategic Cropping Land' is being identified[18] so that it can be protected against any mining development that would cause it to be permanently alienated.[19] The government in NSW is proposing a policy that would see a tighter approvals process implemented where mining is to be carried out in agricultural land.[20] The Queensland legislation effectively gives agriculture 'trump status' when it is being carried out in areas designated as 'strategic cropping land'. This is even the case when the economic surplus derived from mining is far in excess of that derived from agriculture and when, as a result, the bid price for agricultural land offered by mining interests is well in excess of the going market price for the land as farming country.

With such relatively high prices available for their land, why would farming interests be so reluctant to respond by selling their (relatively abundant and hence cheap) land for use in the production of (relatively rare and hence expensive) coal? At least part of the answer to that question is to do with property rights. While the surface land rights

---

18 This raises the question of how a government goes about identifying such lands. Do public officials have the necessary information to determine the best future use of land given constantly evolving production technologies and shifting product demands? With no market incentive to 'get it right' the answer is 'probably not'.
19 http://www.derm.qld.gov.au/land/planning/strategic-cropping/index.html
20 http://nsw.nationals.org.au/Latest-News/nsw-liberals-a-nationals-announce-strategic-regional-land-use-policy.html

to an area proposed for actual mining operations are well-defined and traded in competitive markets, other resources used by neighbouring farmers are not so well-defined. These include the surrounding air space (relevant for noise and views but especially dust that can contaminate off-site pasture and crops) and perhaps most importantly, both surface and underground water. Where the mining operation causes deteriorations in water quantity and/or quality, off-site farmers may be harmed. The rights to water, in particular the aquifers, are not so well-defined. Mining interests are therefore not required to pay for these neighbours' lost resource access.

Establishing, at reasonable cost, a set of well-defined rights to these water resources, both in terms of quantity and quality, would be a constructive step toward establishing a trade-off between all three interests that is more in society's best interest.

Another example of a three-way trade-off is the development of gas reserves that are located in coal seam and shale geological formations through the process known as hydraulic fracturing or 'fracking'. The process involves a fluid being injected at high pressure into the rock strata carrying the gas. The ensuing fractures in the rock formation allow the gas to flow out of the otherwise impermeable rock back to the well-head.

In this case, the trade-off conflict between mining and agriculture/the environment caused by fracking is not so much over land. The surface 'footprint' of the mining operation is relatively limited and does not significantly compromise either agricultural output or the provision of environmental goods and services. However, problems of the 'win–lose' type emerge with respect to the process's impact on the underground water resource. The fracturing of rock strata can result in dropping water tables, and the use of toxic chemicals in the fracking fluid can reduce the quality of the water in the aquifer. Both of these effects potentially compromise agricultural production and environmental conditions in the vicinity of the well (and beyond).

Just as with the Australian open-cut mining case, agricultural interests around the world affected by fracking have resorted to the political sphere to redress their concerns about water impacts, both at the site of the fracking operation and off-site. In France it has been banned. Governments in South Africa and Australia have both placed fracking under a moratorium.

Working out ways to define and defend rights to underground water (in both quality and quantity dimensions) would again be a productive start to establishing a trade-off between competing interests which ensures that resources are used to their best advantage for society. Making sure that the costs of establishing rights are less than the benefits society would enjoy from the trading opportunities that would emerge is critical to this type of market solution.

A 'little green lie'?

Agriculture and mining are much maligned for their negative impacts on the condition of the environment. Yet everyone is dependent on these industries for their well-being. Clearly there is no appetite for outcomes where no mining or agriculture occurs, just as there is no appetite for a severely degraded environment. The relevant question is where to position society along the continuum that defines the trade-off between mining, agriculture and the environment.

The political process has become an integral part of establishing this trade-off. Governments around the world intervene to encourage environmental protection but they also engage in protection of agricultural and mining interests. In the tug-of-war that is government policy formation, environmental interest groups have an incentive to put forward propositions that portray mining and agriculture as environmental villains. In doing so, they have formed some interesting coalitions. In the international trade negotiation arena, they team with domestic protection interests. In some mining controversies, they even coalesce with farmers.

There is no denying that agriculture and mining do impact the environment. In some cases, their impacts have been devastating. However, the characterisation of miners and farmers as environmental vandals is, or at least should be, an exaggeration and can be classed as a 'little green lie'. This is for two key reasons.

First, the environment is a fundamental input into most agricultural production processes. Farmers therefore have an incentive to care for the environment that sustains their operations. Where farmers do not have ownership rights to the resources they use, this incentive to care is replaced by an incentive to over-use and hence degrade.

Second, environmental degradation is not a primary goal of any mining or farming business. But preventing it can add to costs and may not be done on a voluntary basis when those costs are not at least matched by some benefits. Those benefits may be derived from meeting formal legal responsibilities such as avoiding fines or prison sentences, or informal social responsibilities such as maintaining local support. They may also be derived from market transactions if property rights to resources that produce environmental goods and services can be established in order to form competitive markets.

Hence, ensuring that property rights to the resources used in mining and agriculture are well-defined and defended is the first step to rejecting the agriculture/mining 'little green lie'. With rights in place, market exchange would test the relative strength of the values created by the use of resources in their alternative mining, agricultural or environmental uses.

Admittedly, the costs of establishing and maintaining some resource rights regimes can be so high that the effort is not warranted. For instance, rights to obscure components of biodiversity in a river system that could be affected by spills from the tailings pond of a mine may be costly to define and then defend. However, before resorting to a regulatory solution, which will inevitably lead to political to-ing and fro-ing, and the associated generation of 'little green (and other coloured) lies', testing proposed solutions for their impacts on the well-being of society would be wise.

12: The Precautionary Principle

Proposition: Decisions regarding the future of the environment should be made using the 'Precautionary Principle'.

If there is a risk that a proposed action will harm the environment, the Precautionary Principle requires policy makers to regulate against that harm and to place the burden of proof on those proposing an action that it will not cause environmental damage.

BUT

There is always some risk of environmental harm resulting from human action. Demonstrating that there is no risk of harm is impossible. There are also uncertainties associated with not taking action which the Precautionary Principle ignores.

Better safe than sorry

There are many ways of stating the Precautionary Principle. A number of the concept's definitions are framed as 'double negatives', making them tricky to interpret. For instance, the most commonly cited definition comes from the 1992 Rio Declaration:

> Where there are threats of serious or irreversible damage, lack of full scientific certainty shall not be used for postponing cost effective measures to prevent environmental degradation.

The gist of all that is that where there is some danger of environmental harm arising from a new development, it's best to hold back: Just because we're not sure if the harm will occur is no reason to allow the development to go ahead.

It would seem at first glance that this is simply good sense. The homilies, 'better safe than sorry' and 'look before you leap' come to mind. After all, most of us are cautious when encountering risky situations. We buy cars with seatbelts and airbags in case of a car crash. We take out home insurance policies in case of a fire. We stop our children from playing with sharp knives so they don't cut themselves. Perhaps with such sentiments in mind, the Precautionary Principle has seen widespread application. Much of the climate change regulatory machinery has been established under the Precautionary Principle 'logic' set up by the Kyoto Protocol and the Framework Convention on Climate Change. The Cartagena Biosafety Protocol[1] and the Stockholm Convention on Persistent Organic Pollutants[2] both embody the Principle as an operational requirement. The European Union's legal framework explicitly incorporates the Principle, as do the Hazardous Air Pollutant provisions of the US Clean Air Act and a host of other domestic regulations in numerous countries around the world.

The basic idea – that where future outcomes of current decisions are uncertain and potentially catastrophic we should be cautious about making those decisions and guard against environmental disasters occurring – is indeed wise counsel. This is especially the case where some outcomes are 'irreversible': That is, where making a choice leaves no way of going back. Being cautious in this type of setting necessarily involves the assessment of the benefits as well as the costs associated with making the choice to make the change. Part of this assessment involves understanding the type of uncertainty that is involved in the outcomes. Sometimes, outcomes are not certain but we have an idea of the probabilities with which they will occur from previous experience. In such cases, benefits and costs can be weighted by their

---
1   http://bch.cbd.int/protocol/
2   http://chm.pops.int/default.aspx

probabilities so that 'expected values' of alternative outcomes can be considered. For other outcomes, probabilities will not be immediately evident. But even then, 'best guesses' of probabilities can be taken and the decision's sensitivity to different guesses can be judged. Especially where irreversible outcomes are involved, the option of not making an immediate choice so that more time can be taken to understand better the prospects of change may be most attractive.[3]

This process of making decisions when outcomes are uncertain is commonplace in our personal choices and in the assessment of investment options made in the corporate world. What then sets apart the Precautionary Principle as anything special? The answer to this question lies in the subjectivity of the wording of the Principle.[4] Specifically, it can be interpreted in a weak form, much as it has been to date in this chapter, or it can be taken in a much stronger form.

The strong form of the Precautionary Principle calls for restrictions to be placed on options where there is the possibility of environmental harm arising, no matter what the probability is of the harm occurring and no matter what the cost that may be the result of imposing those restrictions. Effectively, the Principle asks us to ignore the costs of the restriction. There are two points associated with this strong version. The first is that the burden of proof is placed on the advocates of change to demonstrate that no environmental harm will occur under their proposal. The second is that the process of implementing the

---

3 Economists have called this benefit arising from waiting for more information to become available the 'quasi option value'. See: Bishop, R. (1982), 'Option Value: An exposition and extension', *Land Economics*, **58** (1): 1–15.

4 Even though the Principle lacks a 'definitive formulation', it has been transformed into a binding legal rule so that its application has become mandatory. Not even precedent has been able to secure a firm definition of the Principle. Within the European Union, applications have ranged from it being used to give absolute 'trump status' to the environment right through to it being used 'at the margin' to tip the balance of evidence toward the environmental protection outcome. See Marchant, G. (2003), 'From General Policy to Legal Rule: Aspirations and limitations of the precautionary principle', *Environmental Health Perspectives*, **11** (4): 1799–1803.

Precautionary Principle should be inclusive, involving all those who are affected by the proposed change.

Weak vs strong

The weak form of the Precautionary Principle allows decision-making to progress through an analysis of the trade-offs that are involved: Are the expected benefits of the proposed change greater than the expected costs? There is an explicit recognition that any change may involve costs, be they environmental or other costs. But it also acknowledges that the change is also expected to generate benefits. The strong form gives trump status to the environment: Wherever the chance of an environmental cost is anticipated – no matter how vague the scientific understanding of that harm may be – the change will be disallowed. This is to be the case no matter what the expected benefit of the change may be.

For instance, the Wingspread Statement[5] defines the Precautionary Principle in the following terms: 'When an activity raises threats of harms to human health or the environment, precautionary measures should be taken, even if some cause and effect relationships are not fully established scientifically'. While the Rio Declaration definition moderates the Principle for application when risks are 'serious and irreversible', the Wingspread version applies to any risk.

Furthermore, under the strong version, the proponents of change have to demonstrate that there is no chance of environmental harm arising from their proposal under the burden-of-proof assignment that can be characterised as 'guilty until proven innocent'.

The implications of these two features of the Principle are profound.

First, the prospects of any proposal for change being able to satisfy the condition of zero chance of harm are slim. It is hard to imagine

---

5 This version – named for the conference centre in Wisconsin where it was developed – was drafted by the 'treaty negotiators, activists, scholars and scientists from the United States, Canada and Europe' with the goal of furthering the application of the Precautionary Principle: http://www.sehn.org/wing.html

a development that does not carry some chance of imposing some environmental harm. For instance, looking back to the introduction of widely implemented developments, have any not had some environmental impact? We all enjoy the mobility afforded by cars and planes, yet they are the sources of a variety of local, regional and global air pollutants that give rise to a variety of negative impacts ranging from bronchial disorders, reduced visibility and (potentially) global climate change. High voltage electricity transmission lines give cause for concern about their emission of radiation, as do microwave ovens and mobile phones. Even the damming of rivers to provide drinking water supplies to cities has caused damage to riverine ecosystems and loss of biodiversity.

All of these developments – cars, electricity and reticulated water supplies – have been introduced because of the benefits they have generated and despite their costs. Put simply, the total benefits to society outweigh the total costs to society of going ahead with the developments. An integral part of that decision logic is that society has been willing to bear the environmental costs associated with them in order to enjoy their benefits. By putting a barrier in front of development proposals, the Principle denies society access to the potential benefits that developments bring.

Placing the burden of proof on the proponent makes doubly sure that proposals for change will not advance where the strong version of the Precautionary Principle is invoked. The impossibility of a proponent being able to guarantee no future environmental harm is apparent. That is the nature of uncertainty. For instance, when organochlorine pesticides were introduced, they were considered to be miracles of progress. Their use allowed the control of malarial mosquitoes as well as crop-destroying pests. People around the world – often those in developing countries – were made better off. Yet no-one could have anticipated the bioaccumulation problem that would eventually cause organochlorine chemicals such as DDT to be withdrawn from use.[6] The consideration of the trade-off between benefit and cost changed with the passage of time and the development of knowledge. But if organochlorines had not been introduced, their contribution to people's well-being would have been lost and the experience generated through their development and use would not have been there to provide the foundations for the introduction of the next generations of less harmful substitute chemicals.

Put simply, nothing can be known with certainty about the future. Requiring proof of no environmental harm will guarantee the failure of any proposal for change.

Application asymmetry

So the application of the strong version of the Precautionary Principle is bound to prevent change occurring. The irony of the Principle is, however, that it will also prevent the continuation of the *status quo*.

If 'no change' is taken to be one possible future option, can it pass the Precautionary Principle test? Can the *status quo* be guaranteed not to result in any chance of environmental harm – and even irreversible environmental harm – occurring?[7] Can the proponents of 'no change' satisfy the burden of proof that there will be no possibility of environmental harm? The answer to all three questions is 'no'. Just as the Precautionary Principle provides a barrier to change, so too does

---

[6] Similarly, the introduction of chlorofluorocarbons (CFCs) as refrigerants and aerosol propellants was not predicted to cause depletion of the stratospheric ozone layer. Nor was the mining and use of asbestos known to cause mesothelioma. These were all cases of ignorance, rather than uncertainty, and so not covered by the Precautionary Principle.

[7] The case of the introduction of the motor vehicle illustrates this point. While cars are certainly the source of a range of pollutants, the alternative which it replaced also had negative consequences for the environment. Not only did horse manure pose a significant waste disposal problem for urban communities, the demand for horse feed increased the amount of natural ecosystems converted to agriculture. The environmental consequences of replacing cars with horses now would be catastrophic.

it stop the continuation of the *status quo*. The example of genetically modified *Bacillus Thuringiensis* (BT) cotton serves as an example.

The introduction of a cotton variety that was able to resist insect attack allowed farmers to reduce their use of insecticides that had damaging impacts on the environment as well as human health. Growing BT cotton would not pass the Precautionary Principle test because it is impossible to state with complete certainty that it will not cause environmental harm, perhaps via the 'gene-jumping' process whereby the genetic modifications made to the cotton plant are inadvertently transmitted to other plants.

But the option of continuing to use pesticides to control the boll weevil also involves (known) environmental harm, so it too would be inadmissible.

What options then are left? Perhaps to grow cotton but without the use of any pesticides or GM varieties? This option too involves risks and even unknown outcomes. They may not be environmental risks but they are risks all the same. Cotton farmers would experience reduced incomes due to the loss of crops from insect attack. Some farmers in developing countries might face starvation. Indirect environmental impacts could also arise. With lower incomes, farming practices might start to induce more land and water degradation: The struggle to survive may mean farmers crop in desperation for an income rather than with long-term sustainability in mind.

What the Precautionary Principle leaves us with, if it is applied symmetrically with all risks being considered in the same way, is a formula that leaves us in limbo. It fails to give any guidance as to how decisions can be made.

Yet its application in its strong form is used as a decision-making framework that prevents development from occurring. Only when applied asymmetrically (so that risks to the environment are the only ones taken into account) can it provide decision rules, but then its role is to protect the environment at any cost.

Some formats of the Principle even stifle scientific research endeavours. For instance, where the 'precautions' defined under the Principle involve the implementation of best available technology (BAT) to control a hazard, the incentives to improve that technology are stymied: With the regulatory condition satisfied, those responsible for the hazard will have no further incentive to do better. The US Clean Air Act provides an example of this perverse outcome. Since 1990 when the Act was passed, there has been a marked reduction in the extent of research into Hazardous Air Pollutants (HAPs).[8]

A problem of definition

Being critical of the strong form of the Precautionary Principle may be seen as an attack on a 'straw man'. If the weak version of the Principle is soundly based in conventional analysis of decision-making under conditions of risk and uncertainty, then what is so special about the strong version? This would appear to be a matter of defining the conditions under which the strength of the Precautionary Principle is ramped up from 'weak' to 'strong'.

Two key elements of these conditions are apparent. The first is concerned with the 'threshold' level of environmental harm and the level of risk that the harm will occur. The Precautionary Principle, in its strong form, is invoked once the environmental harm that may occur as a result of a change becomes 'significant' or 'serious'. This threshold level of harm must be defined. Then the risk that the threshold will be exceeded must also pass some threshold point for the Principle to be invoked. This too must be defined.

The second concerns the extent of the costs that are caused by the choice to invoke the Principle. If the Principle is invoked, then a proposed change is prevented from being implemented. That means the benefits of the change will be lost. That is a cost – what economists

---

8 Goldstein, B. and R. Carruth (2005), 'Implications of the Precautionary Principle: Is it a threat to science', *Human and Ecological Risk Assessment*, **11**: 209–219.

call an 'opportunity cost' because it relates to giving up the opportunity to enjoy a benefit. A caveat to the operation of the Principle is that if those costs are sufficiently 'significant', then it should not be invoked.[9] Hence if an environmentally harmful development can be shown to provide society with a particularly large benefit, the costs of forgoing the development may be deemed sufficiently high to prevent the Principle from being invoked. For example, the construction of a new dam in a developing country may provide flood mitigation, irrigation and hydro-electricity benefits that are so impressive that the operation of the Principle may be suspended. This could be despite the dam's construction causing the loss of an endangered fish species in the flooded river valley.

With these definitional points in place, decision-making is 'business as usual' until the risk and the environmental harm become 'significant' or 'serious' and when the costs of preventing the harm are not 'significant'. Then the Precautionary Principle comes into play and implementation of the proposed change is either prevented or regulated. For instance, in the dam construction example, if the risk of causing a fish species to become extinct is judged to be 'minor' the Principle would not be invoked. If the risk was 'significant', then the Principle would be invoked only if the benefits of the dam's construction were 'not significant'.

The critical question then becomes just what is meant by 'significant'. How is significance determined and by whom? There is no guidance to answering these questions in the Precautionary Principle itself. Rather, the answers are left open to interpretation. When this is the case, the key issue is who gets to determine what is and isn't 'significant'.

---

9 Similarities between the Precautionary Principle and the notion of the Safe Minimum Standard can be drawn: Bishop, R. (1978), 'Endangered Species and Uncertainty: The economics of a safe minimum standard', *American Journal of Agricultural Economics*, **60**: 10–18.

This brings us back to the second point associated with the definition of the strong version of the Precautionary Principle noted earlier in this chapter: 'The process of implementing the Principle should be inclusive, involving all those who are affected by the proposed change'. Essentially this defines a political process in which people with an interest in the proposed change participate. Whether or not such a process will deliver outcomes that are in the best interests of society as a whole is debatable. The prospect is for politically active groups to 'capture' the process for their own specific purposes. In particular, where the costs of preventing change from occurring are spread across the whole population but the benefits of preventing environmental harm are keenly sought by a small and vocal group, the process could well result in the invocation of the Precautionary Principle so that, overall, social harm is caused. In other words, the expected costs of stopping the proposal are greater than the expected environmental harm that was so prevented.

The definition of significance, therefore, becomes one that can be interpreted to serve the interests of those particularly interested in protecting the environment against harm.

The introduction of stringent restrictions on growing genetically modified organisms (GMOs) in Europe[10] provides an example. A key feature of GMOs is their capacity to increase the productivity of agriculture. Restricting their use would result in consumers paying higher prices for food: Without the production of GM crops, the supply of food would be less bountiful and with the same level of demand, prices would rise. But GMOs are also keenly sought after by farmers wanting to increase their profitability – that is, if those farmers are not already protected against import competition.

For European farmers, the use of GMOs only acts to increase the competitive disadvantage they face on world markets, as they allow developing country farmers to produce at even lower cost. It is

---

10 http://ec.europa.eu/food/food/biotechnology/index_en.htm

therefore in their best interests to try to prevent GMOs from being used on European farms and to stop the importation of GMO foods grown elsewhere. The farm lobby group together with environmental lobby groups who fear the risk of environmental harm being done through the 'escape' of genetically manipulated plant material and those who are concerned about the risks to human health of GMO consumption, make a politically powerful combination. The farming interests may have no particular concerns about environmental harm, but the Precautionary Principle provides them with an opportunity to have their interests furthered alongside those of the environmental and human health concern groups. In comparison, individual food consumers hardly notice the price increase caused by the reduced supply of food that results from GMO production and importation restrictions: Indeed, consumers may be convinced that non-GMO foods have health and environmental advantages for which they are willing to pay more.

What this illustrates is that the something that is 'significant' to one party may not be to another. Defining 'significance' thus becomes a matter of judgement and those who get to make the judgement call may not do so with the interests of the whole of society in mind. Concentration of political power helps to predict when the Precautionary Principle will be invoked but, on occasion, the bureaucratic processes involved can also come into play.

An illustration of this comes from the health area. The introduction of new drugs involves risks of harmful side-effects that may arise only after prolonged use of the drug or with a substantial time lag.[11] To guard against the introduction of a drug that eventually produces dangerous side-effects, extensive testing protocols have been developed that can delay the introduction of a new drug for many years. The incentive for

---

11 Perhaps the most infamous case was the morning sickness drug thalidomide, prescribed to women in the early stages of pregnancy. Children born to these women were found to have an abnormally high proportion of birth defects. See: http://www.time.com/time/magazine/article/0,9171,873697-1,00.html

using the Precautionary Principle as a rationale for drug testing comes from those in the government agencies who are responsible for their introduction.

For a bureaucrat to allow a drug to be put on the market that later is shown to be the cause of serious side-effects would be a career-ending move. In contrast, delaying its introduction is low-cost to the bureaucrat. He or she therefore has a strong incentive to be averse to the risk of allowing the drug's release. Again, the Precautionary Principle provides a useful rationale for this risk-averse behaviour. Similarly, politicians responsible for these choices would find their decision to release a drug that proved problematic particularly regrettable from a re-election perspective.

Delays may, however, mean high costs to others. For instance, the delays mean extensive testing over many years. This is a costly process that adds to the costs of the drugs that are finally allowed onto the market. Future users of the drug therefore pay more for them.

An even higher cost may be borne by those who are looking to the newly developed drug as a means of alleviating their chronic symptoms or perhaps allowing their survival. For these people, the potential risk of side-effects from the use of the drug may be insignificant compared with the consequences of not being able to take it.

A 'little green lie'?

Two key characteristics of 'little green lies' are evident in the operation of the Precautionary Principle. First, it is focused on one dimension of an environmental issue. The strong version of the Principle completely ignores the costs associated with protecting the environment. Furthermore, it ignores the environmental (and other risks) associated with the continuation of the *status quo*.

Second, this focus on a single dimension reflects the preferences of those who advocate the Principle. Their concern for the condition of the environment is such that the other aspects of the choice to make a change are relatively minor. This is a view that is not necessarily shared by the rest of society. For instance, a miner's family may be more concerned about the risk of lost employment than the environmental risks posed by the extension of a mining operation.

Implementing the Principle leaves society in a paralysed position, unable to take advantage of albeit risky possibilities of future gains in prosperity and environmental improvements. This paralysis is worsened if the Principle is applied 'even-handedly' across different types of risks to different types of benefits and costs. The future is unknown not just with regard to environmental impacts. Imperfect knowledge is ubiquitous. There will never be complete scientific certainty about the consequences of any course of future action. Lacking such certainty should therefore not be used as a reason for or against any particular choice.

The popularity of the Precautionary Principle can be explained by its usefulness to those with vested interests. These are not just those with environmental protection interests but also those who may be able to gain through coalition with environmental interests. These include labour unions and farmers seeking protection from market competition.

Armed with an apparently logical concept, advocates of environmental protection have been able to convince politicians that there is merit in 'looking before you leap'. Politicians and the bureaucrats working alongside them, find caution appealing. Making a decision that proved to have undesirable outcomes would be a personal disaster while options characterised by 'safety first' can be relatively easily 'sold' to the voting public. This is especially the case when lobby groups may be running publicity campaigns that display the 'dire consequences' of change. The media is well attuned to publicising stories involving potential environmental disasters given the adage that 'good news never made a paper sell'. And the general public is particularly vulnerable to these bad news stories.[12] Sunstein[13] suggests a range of reasons for this vulnerability:

- Loss aversion. People put more weight on losses than equivalent gains.
- The myth of a benevolent nature. Natural processes are believed to be 'safer' than human interventions.
- The availability heuristic. Risks that are prominent (potentially through being publicised) are seized upon, while less salient risks are ignored.
- Probability neglect. The probability of a bad outcome occurring is given little attention compared with the magnitude of that outcome.
- System neglect. People fail to recognise the wider implications of change while focusing on a specific outcome.

The Precautionary Principle thus becomes a convenient mechanism for 'little green lie' advocates to 'protect' people from harming themselves by accepting environmentally damaging changes. However, its implementation brings with it the dangers of 'hobbling' society in its quest for improvement. It would be better if changes were analysed – giving due recognition to risks and uncertainties of all kinds – by weighing up the relative merits of the options available. Better still that this process of selecting future directions for society be guided by the exchange in markets of well-defined property rights across the resources involved. Such market exchange deals with risk. If there is a chance that consumers of a good will bear some harm, demand for

---

12 An infamous non-environmental example was the Y2K 'scare' by which the world's computers were predicted to collapse on the change in year from 1999 to 2000. While some people made fortunes designing and installing software 'solutions', many bore the costs of what turned out to be unwarranted action. Health risks, especially those relating to cancer, reproductive health and heart disease, are a continuous source of excessive concern, be it over eggs, sugar, peanuts, grilled meat, etc. Environmental 'false positives' where over-regulation has occurred, have included genetically modified crops and numerous agricultural herbicides and pesticides.
13 Sunstein, C. (2002), 'The Paralyzing Principle', *Regulation*, Winter: 32–37.

the product will fall and less will be supplied. Producers will seek ways of satisfying consumers in ways that reduce risk so that demand for their products is increased. It is through the process of exchange that mutually beneficial outcomes will be explored and seized.

Index

A
abatement, 163, 177–178, 181–183, 212
accreditation schemes, 54, 56, 61
adaptation, 75–77, 82, 164–165, 183–184
agricultural productivity, 74, 211–212, 214
agricultural protectionism, 221
agriculture and mining, xxi, 209–231
air pollutants, 33, 234, 237, 240
alternative energy sources, 15, 19, 25
alternatives, 7–8, 12, 14–16, 18–19, 23, 34, 57, 64, 75, 98, 118, 123, 144, 183, 194–198
asset value, 37–38, 199
Australian Conservation Foundation, 203, 215
Australian Wildlife Conservancy (AWC), 199–200, 202

B
Bacillus thuringiensis, 214, 239
Baptist and the bootlegger, 65–66, 107
Basel Convention, 104–105, 108
benefit–cost analysis, 108, 157, 160–161, 171, 173–174, 184
bio-diesel, 29
biodiversity loss, xxi, 209
biofuel, 21, 36, 177
birth control, 70, 77
Borlaug, N., 214
BP, 5
Bruntland Commission, 50
BT cotton, 214, 239
Bush Heritage Australia, 200

C
cap-and-trade, 40, 195
carbon credits, 166, 175
carbon dioxide, 33, 50, 167, 213
carbon footprint, xvii, 43, 47, 50, 53, 65
carbon tax, 163–164, 178, 182
Cartagena Biosafety Protocol, 234
cartel, 9
Choice Modelling (CM), 156–158, 161
choke price, 11–12
chronometer, 89–90
Ciriacy-Wantrup, S.V., 153
CITES, 204–205
clean energy, xvii, 21

climate change, xv, xx, 23, 33, 47–48, 54, 163–184, 222, 234, 237
Club of Rome, 17
coal, 7–9, 22–24, 30–31, 33–35, 37, 40, 68, 93–95, 97–98, 119, 157, 177, 180, 183, 227–228
coal-fired thermal power, 22
collective action, 40, 150, 194–199
Common Agricultural Policy, 221
Commonwealth Environmental Water Holder (CEWH), 203
comparative advantage, 22, 57, 101–104, 106–107, 182
consensus, xv, 166–167
Conservation Reserve Program, 36
consumption index, 45
costs, ix, xi, xii, xiii, xix, xx, 6–7, 11–16, 19, 22–23, 25–26, 28–39
crowding out, 118, 172, 193, 203

D
DDT, 238
discounting, 118–119, 170–174
donations, 154, 198–203
dumping, 115–116, 122–124, 127

E
Earth Sanctuaries Pty Ltd (ESL), 199, 202
ecological footprint, xvii, 43, 49–50, 53, 56–57, 59, 64
economic efficiency, 132–133, 141, 178
economic growth, xiv, xviii, 4, 12, 22, 80, 82, 84, 86, 89, 92–97, 100–102, 105–106
efficiency, xix, 8, 12–13, 18, 27, 101, 129–134, 136, 139–144, 178
efficiency measures, xix, 129, 134
Ehrlich, P., 17, 68, 73, 75–76, 86, 152
embodied energy, xvii, 43, 46–47
endangered species, 36, 94, 99, 188–189, 201, 204, 223, 241
endosulphan, 213–214
energy rating, 138
environmental amenity, 125, 127
environmental decay, 93, 101, 107
environmental degradation, xxi, 93, 95, 99–100, 190, 209, 230, 233
environmental protection, xii, xviii, 62, 64, 89, 101, 140, 187, 190–193, 202, 204, 206–207, 210, 216, 218, 223, 229, 235, 245
enviropreneur, 204
equity, 27, 92, 94, 208
ethanol, 29, 31–32, 35–36, 177
EvFit, 69, 71
EVRI, 158
exchange, 12, 25–26, 28, 57, 61, 73, 149, 193–199, 230, 246–247
exploration, xvi, 1, 5–6, 10, 193
external institutions, 194
extractive industries, xxi, 209
Exxon Valdez, 33

F
fair trade index, 48, 53–55, 61
Fairtrade International, 48
false economy, xvii, xix, 43, 129

Farm Bill, 221
fertility rate, 83–85
food crisis, 32
food miles, x, xi, xvii, 43, 48, 53–54, 65
forced sterilisation, 70
Forest Stewardship Council, 49
fossil fuels, xii, 8, 12, 19, 22–26, 33, 35–36, 40, 69, 73, 165, 183, 213, 222
fracking, 228–229
free-riding, 177
fuel efficiency, 13, 18
Fukushima, 23
future generations, xvii, xviii, xix, 27, 43, 67, 89, 117, 129, 172–173, 179

G
Garnaut Review, 170
gas, 7–8, 11, 22–24, 177, 228
geothermal, 12, 21–22, 24
GHG emissions, xx, 163, 174
global climate change, xv, xx, 33, 48, 163, 237
global financial crisis, 4, 7, 11, 32
Global Footprint Network, 49–50
global village, 90
global warming, xv, 165, 167, 180
globalisation, 89–91
GM, 213–214, 239, 242
Gore, A., 165
Green Revolution, 214
greenhouse, 33, 35, 47, 65, 130, 140–141

greenhouse gas, xx, 33, 35, 47, 65, 130, 136, 140–141, 163–164, 166–171, 174–176, 178, 180, 222

H
Hardin, G., 187, 196
hazardous waste, 104–105
Healthy Rivers Australia, 202–203
Hotelling, H., 10, 38, 119
Household Responsibility System, 72, 74, 76
human capital, 74, 76, 82, 97–98, 173, 212
hybrid cars, 141, 144
hydro power, 22, 30, 36

I
incandescent light bulbs, 130, 134, 138, 140
incentives, xiv, xx, 37–38, 72–76, 87, 98, 113, 117, 134, 136, 154, 175, 185, 187–188, 191–193, 200, 205, 207, 217–218, 226, 240
Index, xvii, 43–66
indices, 44–62, 64–65
individually tradable quotas (ITQ), 195, 197
Industrial Revolution, 22
infinite value, xi, xix, 145, 147, 150–151, 158, 210
innovation, 8, 18, 73, 76, 86, 89, 198, 212, 214
interest groups, xiii, 59, 62, 106–107, 142, 144, 160, 180–181, 219, 223, 229

interest rate, 10, 119, 139, 171–173
internal institutions, 196
international trade, xviii, 12, 57, 72, 89, 101–102, 204, 223, 229
IPCC, 166–168, 170, 174
irreversible, 85, 95, 100, 152, 212, 227, 233–236, 238

J
joint goods, 201, 205

K
Kuznets Curve, 94–95, 97–98, 100, 102
Kyoto Protocol, 166, 174–177, 183, 234

L
Landcare, 215
landfill, xviii–xix, 109–117, 121–122, 124–125, 127, 141
Life Cycle Analysis, 46, 57
life-cycle assessment (LCA), 51–53, 59, 65
liquefied petroleum gas, 7
litter, 115, 123–125
little green lie, x–xx, xxii, 16–19, 23, 37, 41–43, 56, 64–66, 85–87, 92, 105–108, 126–127, 142–144, 158–161, 180–184, 207–208, 210, 229–231, 244–247
lobby groups, 39, 42, 96, 159, 203, 215, 218, 222–224, 243, 245
Lomborg, B., 96, 174
loss aversion, 246

M
Malthus, T., 68
manufactured capital, 74, 97–98, 173, 182
market based instruments, 207
market exchange, 12, 26, 28, 149, 195, 230, 246
market failure, 187, 190–191, 193, 207
market forces, 9, 40, 190, 207, 218
markets, xv, 25–28, 31–32, 38, 41, 61, 76, 80, 82–83, 86–87, 101, 115, 132, 134, 137, 144, 148–151, 153–154
Middle East, 9, 11
mitigation, 163, 165–166, 170, 178, 180
mitigation policy, 164, 169, 171–175, 183, 241
monopoly, 9
monopsony, 9
Montreal Protocol, 183
Multi-Criteria Analysis (MCA), 51–52, 58, 160

N
National Parks, 94, 99, 150, 156, 187–188, 198, 206
natural capital, 97, 173
natural gas, 18, 35, 41
NIMBY, 110
nitrous oxides, 23, 33, 97
non-renewable energy, xiii, xvii, 21, 23, 27, 29, 34, 40–41, 43, 92
non-renewables, xii, xiii, xvii, 5,

21–23, 27, 29, 31, 34–35, 37–38, 40–42, 44, 69, 118, 212
not-for-profits, 193, 198, 200–203, 208, 225
nuclear power, 12, 23–24, 73–74

O
OECD, 4, 151
offset, xxi, 5, 29, 40, 181, 189–190, 209
oil, xvi, xxi, 1–19, 22–24, 27, 31, 33–34
oil 'fuel gauge', 2
oil consumption, xvi, 1, 3–5
oil reserves, 2, 4–6, 12, 16
oil shortages, 13
oil substitutes, 7, 16
one-child policy, 70–71, 84
OPEC, 9, 12
open-access, 187
open-access resources, 187
open-cut mining, 226, 229
opportunity cost, 55, 79, 81, 84, 173, 241
'organic' certification, 48
Ostrom, E., 196, 199
ozone, 33, 35, 183, 238

P
payments for environmental services (PES), 189
peak oil, xvi, 1–19, 27, 41, 45
petroleum reserves, xvi, 1
picking winners, 39
plastic bags, 125–126, 140

PM10, 95, 102
pollution, x, xii, xiii, xiv, xviii, xix, 27, 32–36, 40, 44–45, 58, 93, 95, 99–100, 103, 107, 109, 111, 150, 154, 158, 187, 195, 197, 206
population, xiii, xiv, xviii, 49, 67–87, 92, 103, 106, 142–143, 146, 173, 179, 210, 219–220, 242
population choices, 77
population controls, 70–73, 82, 84, 86–87
population decline, 85
population growth, xviii, 67–69, 71, 76, 81, 83–86
Precautionary Principle, 240
price of oil, 6–13, 16
price rises, xvi, 1, 10, 38, 75, 118
price signals, 17, 61, 63, 120, 134, 136–137, 143, 221, 224, 226
prices, xvii, 8–15, 18, 28–29, 31–32, 34, 37–39, 43, 48, 55, 60–64, 66
'principal–agent' problem, 161
private property rights, 26, 74–75, 98, 189, 218
private sector, xx, 41, 99, 172, 185, 190–195, 197–199, 201–204, 206–208
production subsidies, 14–15
productivity, 23, 72, 74–76, 97–98, 129, 177, 211–212, 214–215, 225, 242
Property and Environment Research Centre, 204
property rights, 26, 33–34, 61, 74–

75, 80, 82–83, 86–87, 98, 114–117, 120–122, 132, 134, 137, 149–150, 186–187, 189, 193–196, 217–218, 225–227, 230, 246
protection, 198–201
protectionist policies, 221–223
public goods, xx, 87, 99, 101, 123, 158, 185–188, 190–191, 193, 195, 201, 205, 207
public sector, xx, 80, 99, 172, 185, 190, 193–195, 197–198, 201–203, 207–208
public/private partnerships, 206

Q
quotas, 15, 18, 75, 91, 99, 106, 195, 208

R
rationing, 13–15, 17–18, 63–64
real price, 11
rebound' effect, 142
recycling, xviii, xix, **18, 104–105, 109–110, 117–121, 125–127, 141, 157**
regulatory impact analysis, 150
relative scarcity, 10, 31–32, 37–38, 41, 60–61, 63, 87, 102, 119, 125, 132, 136, 149, 225
renewable energy, xii, xiii, xvii, 19, 21–42, 45, 65, 99, 165
renewable energy certificates, 29
rent-seeking, 17, 224
reserves-to-production ratio (R/P) **ratio), 5–6, 8–9, 19**

resource allocation, 25, 27–28, 62, 83, 222
Rio Declaration, 233, 236
risk, xx, xxi, 23, 33–34, 41, 53, 56, 59, 70, 78, 80, 83, 124, 139, 142–143, 159, 161, 163, 167, 171, 177, 180, 183–184, 186, 214, 225, 233–234, 236, 239–241, 243–247

S
safe minimum standard, 153, 241
scarce resources, xiv, xvii, xviii, xix, 14, 17, 30, 43–47, 49, 59, 63, 67–68, 89, 91, 102, 104–105, 109, 114, 120, 122, 126, 129, 131–132, 139, 141, 143, 184, 186, 193–194, 196–197, 221
scarcity, xvi, xvii, xviii, 1–3, 10, 31–32, 37–39, 41, 43–44, 47, 50–53, 59–65, 67, 69, 74–77, 82, 86–87, 92, 96, 114–115, 118–120, 125, 129, 132, 134, 136–137, 143, 149–150, 224–225
scarcity signalling role, 61–62
sceptics, 167
scientific method, 167
shrimp-turtle case, 223–224
Simon, J., 17, 73, 75–76, 82, 86, 94
social capital, 74, 76, 97–98
soil erosion, xxi, 51, 82, 98, 188, 209, 211, 213–214, 220, 226
solar panels, 12, 19, 39–40
Solyndra, 40
sovereign risk, 225
special interests, xii

specialisation, 91, 101, 182
species extinction, 100, 152
species loss, 95
starvation, xviii, 41, 67–68, 72, 83, 239
stated preference, 151–152, 154–156, 158–161
Stern Review, 170
Stockholm Convention on Persistent Organic Pollutants, 234
Strategic Cropping Land, 227
Subsidies or subsidy, 14–15, 18–19, 28–29, 36, 39–40, 42, 62, 103, 117–118, 124, 127, 130, 133–134, 140–144, 150, 157, 177, 181, 205, 207, 216, 219, 221–222
substitute, xvi, 7, 9, 11–12, 16–18, 31, 37–38, 55, 58, 80, 117–119, 127, 129, 131, 133, 136–137, 143–144, 170, 183, 238
substitute energy, xvi, 1
sulphur dioxide, 23, 33, 35, 94–95
supply and demand, 4, 15, 148
sustainability, 48, 50, 55, 64, 92, 190, 239
sustainable population, 69

T
Tall Green Tales, xxii
TANSTAAFL, 161
tar sands, 8, 11, 33
tariffs, 91, 106, 221
technical efficiency, 131–133, 144
technological advances, xvi, 1, 22, 63, 73, 132

The Nature Conservancy, 199
threshold, 85, 100, 124, 152–153, 211, 240
tidal plants, 12
time discounting, 118
Tinbergen Principle, 41, 101
trade, xviii, 12, 22, 26, 28, 57, 62, 72, 75, 89–108
trade-off, xxi, 55, 58, 96–97, 143, 149, 151–156, 158–159, 161, 209–212, 214, 216–219, 224–226, 228–229, 236, 238
transaction costs, 26, 91, 136, 150, 194–198, 206–207
triple bottom line, 50–51

U
unintended consequences, 71, 105, 108, 117
US Clean Air Act, 234, 240
use value, 26, 37, 60
usury, 90

V
vested interests, xv, 18, 58–59, 62, 107, 165, 180, 245
virgin material, 117–120
virtual water, xvii, 43–47, 49
visual pollution, 36
Voluntary Human Extinction Movement, 71, 146

W
waste, xviii, xix, 24, 47, 49, 93, 103–105, 107, 109–127, 140, 145,

178, 209, 212, 215, 238
waste hierarchy, 110–111
waste management, 117
waste reduction, 117–121, 126–127, 140
water footprint, 44–45
water rights, 202
weighted indices, 58
willingness to pay, 26, 151, 153–155, 179
win–win, 97, 211–217, 225
wind generators, 12
Wingspread Statement, 236
world economy, 7, 11, 32
World Meteorological Organization, 166
world population, xviii, 67–69, 71, 73, 83
World Trade Organisation (WTO), 91, 106, 223–224
worldometers, 2, 5, 8–9, 68

Z
zero tillage, 213
Zero Waste, 110–111, 114, 117

www.ingramcontent.com/pod-product-compliance
Lightning Source LLC
Chambersburg PA
CBHW031725230426
43669CB00007B/240